D0906089

Generation Dread

BY BRITT WRAY

Rise of the Necrofauna: The Science, Ethics, and Risks of De-Extinction

Generation Dread

Finding Purpose in an Age of Climate Crisis

BRITT WRAY, PhD

ALFRED A. KNOPF CANADA

PUBLISHED BY ALFRED A. KNOPF CANADA

 Published in 2022 by Alfred A. Knopf Canada, a division of Penguin Random House Canada Limited, Toronto. Distributed in Canada and the United States of America by Penguin Random House Canada Limited, Toronto.

www.penguinrandomhouse.ca

The author is not clinically trained and none of the content in this book is clinical advice. The author is writing as a researcher and science communicator, in the hopes that what this book contains is helpful for readers.

Library and Archives Canada Cataloguing in Publication

Title: Generation dread : finding purpose in an age of climate crisis / Britt Wray.
Names: Wray, Britt, author.
Identifiers: Canadiana (print) 20210261536 | Canadiana (ebook) 20210263482 | ISBN 9780735280724 (hardcover) | ISBN 9780735280731 (EPUB)
Subjects: LCSH: Environmental psychology. | LCSH: Climatic changes—Psychological aspects. | LCSH: Global environmental change—Psychological aspects. | LCSH: Human beings—Effect of climate on. | LCSH: Human ecology—Psychological aspects.
Classification: LCC BF353.5.C55 W73 2022 | DDC 155.9/15—dc23

Jacket images: (leafs, flowers) Balsam Tree (Cenchramidea Arbor) from *The natural history of Carolina, Florida, and the Bahama Islands* (1754) by Mark Catesby (1683-1749). Original from Biodiversity Heritage Library. Digitally enhanced by rawpixel; (yellow and blue butterflies) Illustrations from the book *European Butterflies and Moths* by William Forsell Kirby (1882), a kaleidoscope of fluttering butterflies and caterpillars. Digitally enhanced from our own original plate; (sky) Ben Hershey/ Unsplash; (woman) Katarina Palushaj / EyeEm ' Getty Images

Author photo: © Arden Wray
Jacket design: Kate Sinclair

Printed in Canada

10 9 8 7 6 5 4 3 2

To Atlas, and every soul who is overwhelmed
by this crisis yet refuses to look away.

INTRODUCTION

The emotional journey that comes with facing up to the environmental crisis can be very intense, disturbing, and extremely painful. You'll know this if you have ever cried upon reading that a species you love is going extinct. Or anticipated that you'll soon lose a coastline you call home to the sea. Or witnessed livelihoods shrivel as year-on-year drought choked the land. It is stressful to live in fear of dangerous climate tipping points that, once surpassed, will unleash cascades of self-reinforcing environmental change—from ice sheet disintegration to permafrost loss and dieback of the Amazon—that cannot be stopped or reversed. It is infuriating to learn that this was predicted and preventable but that a handful of powerful figures with entrenched interests knowingly and continuously sacrifice the future and people's well-being for profit. It is haunting to connect the dots between imperialism, colonialism, genocide of Indigenous peoples, racial capitalism, industrialism, and extraction, and to reveal the shadowy footprints of these logics in the places where those who are most vulnerable to environmental traumas now dwell.

Swarms of people are waking up to a sense that, if nothing drastically changes, the fate of humanity will be like the Slinky pushed off the top stair, with no good structures in place to halt its energetic descent. Nothing could be further from what people want, but many feel powerless to stop this from happening, which has whipped up tidal waves of grief, anxiety, pessimism, and existential dread. The worst outcomes are not inevitable, and much can still be healed—but as emotions get bulldozed by world events and scientific predictions, the ability to create a more just and healthy world depends largely on how these difficult feelings are tended to. The young people who must confront the ill state of the planet they've been handed, and the elders whose cries for environmental change have been disregarded for decades, are in an especially frustrating position—but also a potentially powerful one, now that their voices are finally being heeded.

The artists at the Bureau of Linguistical Reality, who are coming up with new language to better describe our changing world, have named a quintessential sentiment of the times *brokenrecordrecord-breaking*: "a recurring feeling of déjà vu, quiet terror, and slow shock which is both acute and familiar that occurs when opening a newspaper, radio program or website and reading a headline that that year (month, season, day) has broken the record for the hottest on record." Many are struggling to stay afloat as we process the steady stream of scary environmental news that tells us things are unravelling even faster than scientists expected. Some lose themselves in activism, while others keep their distance or close their eyes just enough to pretend that the reality we face isn't nearly as bad as it is. For some whole communities who may be contending with the immediate practical threat of hurricanes, heatwaves, floods, rising sea levels, drought, or raging wildfires, closing one's eyes isn't an option.

Increasingly, even for those far away from hazard zones, neither is taking the time to examine one's emotional response.

Over the last few years, especially but not exclusively in liberal circles, the term *eco-anxiety* has become all the rage. It describes a condition that robs sleep from those who, when all is dark and quiet, stir in thoughts of how uninhabitable the Earth will soon become. Tools are cropping up to help people cope with eco- and climate anxiety, grief, and a pervasive sense of powerlessness to halt nature's destruction. Self-care guides, climate-conscious therapists, and a cottage industry of coaches have emerged to help folks grapple with ecological uncertainty, find community support, and focus on the pro-environmental actions they can take. But acknowledging and reckoning with difficult emotions is still not the norm, and mental health resources are cost-prohibitive or simply unavailable for many people who need them.

Too often, though, the conversation around eco-anxiety reveals its own amnesia. Only some of us are being forced to grapple with the threat of annihilation—and the emotional weight this carries—for the very first time. For so many people who've been marginalized, the oppressiveness of how bad things are is a tale as old as the hills. As a white, cis-gendered, economically secure woman, I have the luxury of dreading the future (in light of problems like climate change) while others already acutely fear the present, and have long been suffering for how they were treated by dominant power systems in the past. Unfortunately, the climate crisis creates a double injustice here, as the most marginalized, those who had the least to do with creating this mess—predominantly poor people of colour—are disproportionately harmed by a warming world.

Eco-anxiety researcher Panu Pihkala says that waking up to the climate and wider ecological crisis is particularly hard for middle-class citizens of industrialized nations because "the world is revealed to be much more tragic and fragile than people thought it was." This profound disruption, which can be as severe as an internal shattering, sends them into a grief-stricken process of mourning the lost future they believed would come—a future of ecological stability. This then erodes their sense of security. As more and more people who've been living comfortable lives wake up feeling eco-anxious, that awareness comes with a risk. If we only turn inward, to recognize this pain within ourselves, instead of looking outward, to glean a sense of implication in the far-reaching and unequal consequences of the climate crisis and our agency to improve the outcomes, little will change. Better futures will be entirely missed if we get stuck in fear and dread.

For those who might roll their eyes at the groundswell of anxiety and grief that many privileged people are now expressing, I hear you. We're going to get this wrong if we depoliticize this pain, by not seeing its entanglement with centuries of environmental violence, racism, and domination. Without that context, we cannot be honest about who is the most vulnerable now and going forward, nor figure out how to best reduce harm. On the flip side, the tumultuous feelings that are on the rise are completely valid, need tending to, and present a great opportunity for justice-oriented personal, environmental, and social transformation.

Everyone is vulnerable to the distressing—and potentially revitalizing—power of eco-anxiety, but we don't all have the same resources, space, or interest to harness it when other existential threats may be more immediate. My hope is that by being explicit about these inequalities in *Generation Dread*, this book can contribute to us getting better at looking out for each other, as things get harder and heat up. I am part of this generation, and I know what it feels like to go on a journey of developing critical awareness from one's own existential fears. If you're part of it too, no matter your age—if you want to find inner strength and also help bring about a more sustainable, just, and equitable future—you're among those who can most benefit from reading these pages.

Right, so what is it we're dreading again? It's the fact that we are in the midst of an escalating planetary health crisis. Much more than just climate change, the planetary health crisis is humanity's destruction of nature, and it is affecting everything from the climate to biodiversity, fresh water, fertile soil, clean air, land cover, the spread of infectious disease, rates of chronic disease, and, as a result, the health of all living beings. It is a civilization-changing event that is well under way, causing incredible damage and deepening existing injustices. With each passing day, the realization dawns on more people with even a smidgen of environmental identity—a feeling of connection to the non-human environment—that much of what we love in the world is under threat, and so the collective sense of being traumatized grows. The demand for resources that help people ground themselves and feel capable of creating change is outpacing the supply.

Wealthy nations and elites must act on the climate crisis in order to flip the dangerous trend lines, but most are acting as though they don't understand that to delay is to dance with their own demise. Much as we see with the COVID-19 pandemic, without bold preventive measures in place, few to no rich countries will be spared its devastating effects. Just like our warming climate, pandemics are not separate from, but a symptom of, our planetary health crisis. As we continue sucking resources out of the natural world—by cutting trees in tropical forests, for example, or extracting minerals and fossil fuels—we bump into species that live in the wild places we tear into, and become hosts to their viruses. Epidemiological research shows we can prevent future spillovers and stop outbreaks from turning into pandemics by dramatically changing the way we interact with the natural world.

All this ought to compel us to rethink our relationship to nature itself. As the cries for change along these lines grow louder, denialists clamp down harder, critics police the tone of climate alarmists, climate alarmists burn out while hunting for a sign that their actions are having an impact, and doomsayers make peace with their own death and that of society. "Rational environmentalists," arm in arm with climate dismissives, condemn emotionally charged messaging about the climate emergency with hashtags like #climatecult. Climate scientists and green political leaders quit their jobs and move to the countryside to escape the mental exhaustion of their work. Meanwhile, some people believe that it is already too late to prevent societal collapse. They speak about the "earth as hospice," and suggest we use what time we have left to summon the courage to face the music, as the ship we're all on sinks.

We're in a profoundly turbulent time, and it begs us to build up our wisdom about how we relate to our feelings inside a culture that still values capital over compassion and the well-being of the poor, the "Other," and the not yet born. Unfortunately, emotions have typically been regarded as feminine and a sign of weakness, and so have been undervalued and dismissed. This tendency is deeply harmful, because in order to thrive on a hotter and more hostile planet, we will need a high degree of care, interest in each other's point of view, sensitivity to one another's vulnerability, and patience as we seek sites of commonality. All these qualities will be key to restoring our humanity amidst competing pressures, including tightening borders, rising walls, pointing fingers, and social unrest.

I had a reckoning with my own eco-emotions in 2017 when my husband Sebastian and I thought we might start trying to get pregnant. A deep sense of grief and despair came crashing over me when I considered what it would mean to deliver a child into this world—a world dominated by a small group of greedy humans who are walking with open arms into ecological dead zones, mental breakdowns, and conflict over dwindling resources; humans who won't raise their fists to these calamities because their avarice restrains them.

Sure, we also grappled with the common ambivalences many people list when considering becoming parents (issues of resources, identity, time), but environmental concerns loomed largest. Still, Sebastian and I longed for a child, and felt more drawn to the idea of reproductive alchemy—converting both of our base metals into gold in baby form—than adopting, likely due to several kinds of societal, familial, and biological pressures. This birthed a painful dilemma, from an ethical point of view. When you confront head-on what scientific models say about the suicidal track we're on, alongside the political establishment's completely inadequate efforts to address it, is it okay to decide to bring a new person into this situation?

As a science writer and graduate of a biology program with a focus on conservation, my exposure to research on climate change and the environment was significant enough to put me in that hapless bracket of professionals who are, unsurprisingly, especially stressed by what the science says. I could not resolve my fears by talking them out with Sebastian, or my parents, or my friends. The dilemma had no solution. It was either have a kid, and risk being taxed with crushing anxiety for the rest of my life about how our child will deal with a planetary condition that is becoming deadlier and more devoid of natural wonders; or don't have a kid, and miss out on something we deeply want to do and all the nourishment that kids bring into people's lives. I had to get my heart around this quandary and see if my thinking was perhaps twisted or off base. I started interviewing all kinds of experts, parents, and non-parents, collecting their perspectives like charms on a bracelet that, once full, I imagined would allow me to come to the right decision. As it turns out, dilemmas don't wear jewellery, and no amount of other people's insights could glue what had fragmented inside me back together again. I'd have to be brave enough to intentionally stay child-free, take the leap, or reveal that I was too much of a coward to make a decision, at which point my biology would do my deciding for me.

But then, somewhere between the UN's 2018 Intergovernmental Panel on Climate Change report that outlined the difference between a world at 1.5 and 2 degrees Celsius of warming—which was etched into people's minds as saying *we only have twelve years to avert climate catastrophe*—and the global youth climate strikes of the following year, reproductive anxiety due to the climate crisis had become mainstream. There are plenty of statistics and articles to prove it, which I unpack in chapter 4. That made me wonder how the planetary health crisis might be psychologically affecting people in other ways, including pernicious ones that most of us are not aware of. I started gathering people who know much more than I do about things that fascinate and trouble me—such as

- what's been holding us back from dealing with this crisis head-on;
- how eco-anxiety affects one's thinking and behaviour;
- where the nuances lie in people's concerns around having kids in times of warming;
- what we can do to prevent eco-emotions from becoming overwhelming and debilitating;
- how we can support each other in living more comfortably in the moment while appreciating the immense dangers we face and working to mitigate them;
- what communities that have long lived under existential threat know about surviving dark times;
- and how those with more privilege and protection can supportively partner with them, recognizing that all fates are tied, without re-centring their own needs.

Through conversations, research, and reflection, I learned how we can maintain a balance between hope and fear without giving in—out of self-preservation—to either side of this tempting binary. A crucial part of finding that balance—and sitting in the uncertainty—is grieving what is happening and understanding why mourning ecological losses may hold political power for us now, allowing us to gather the conviction to make change while extending a platform for others to join us. Rather than bury our heads in the sand and suppress our discomfort, we can harness and transform the distress we feel into meaningful actions and forms of connection. It became clear to me that the keys to that transformation are wise communication and building bridges across our divides, as well as supporting young people to deal with the dangers they are inheriting. Lastly, I wanted to understand what happens to mental health when disaster strikes, when people's hope in the future erodes, and what will be required from the mental health field if it is to deal with the scale of psychic damage this is all causing.

Those ideas became the seeds for this book, and what I discovered about them makes up the chapters that follow. They chronicle an education I received by peeling back the layers of many different fields. Partway into my research, I developed an understanding of how my own privileges were fuelling my eco-anxiety. My lack of personal experience with existential threats meant I had a lower reserve of existential resilience to rely on. I'd have to learn to cultivate it through a process of internal emotional activism, which will be explained in the chapters ahead. A key part of that for me involves listening to the stories of people who have not been able to take their security for granted, and who have long understood how unsafe the world can be. What my research clearly shows is that facing your fears and authentically committing to the belief that it isn't too late to strive for better futures is long and exhausting work that requires all forms of care. That's why this book also includes therapeutic concepts for cultivating additional forms of inner resilience from the ones you already hold—because everybody needs support in ontologically threatening times.

In order to shore ourselves up against what's to come, how we organize our society and economy needs to change, how we care for our health needs to change, how we parent needs to change, and how we relate to nature and each other needs to change. Because, whether we like it or not, what it means to be human has already changed. The environment we grew up with is slipping away, making room for the emergence of a new Earth that many scientists have projected in grim detail. But this coming world is not yet fully known—there is still room to shape it. The positive in all this is that the torment comes bearing gifts. If you explore its depths, you'll find a valve somewhere inside you that taps into the most existential part of yourself. Once you open it, a boundless stream of love, connection, and meaning will always be at your back, fuelling what you do. And if you are among those feeling this pain, we cannot afford for you not to also discover this fuel. The planet will be here for a long time, but its humans and many other species might not. They need you to be wakeful, internally strengthened, and externally motivated, and that is not hyperbole.

My own eco-anxiety helped me discover this valve. After reckoning for a couple of years with my fears about what's happening all around us, I jumped into the planetary health field, developed a new specialty in the psychological and emotional impacts of environmental degradation, and refocused my academic research on real-world outcomes for improving young people's mental health and well-being in a warming world. What started as a private preoccupation about whether or not to have a baby developed into a new identity and a much deeper sense of purpose in my work. The way I was led there was not by my brain, but by some mixture of stomach and heart. This is not to pat myself on the back but to point out that eco-emotions can do so much in the world, and we're each being challenged to probe our existential feelings' potential to spur positive action. Research has shown that the kind of psychological distress about the climate crisis I experienced is a strong predictor of behavioural engagement with these problems. Eco-distress needs to rise to the surface and be met head-on for the transformative work to begin.

This book is not a manual for how to take effective climate and environmental action. There are tons of books and resources available on that topic, and they are hugely important and urgently needed. I invite you to seek them out for that kind of practical advice, while I hope *Generation Dread* will help you appreciate that activism is not only external but internal too. The influential figures shaking up how things are done are often driven by emotion; it's a fact that lasting change is propelled by inner clarity, meaning, and purpose.

However you read this book, I hope you will welcome the hurt you are feeling for the world as much as your hope, honouring the role they both play in your life. Be curious about what is inside you, without self-judgment or shame. And if you're feeling numb or even content with the way things are, just be patient with yourself. As you read the following chapters, see if there is anything stirring beneath the veneer of calm. If it makes a sound, get down low and bring your ear to its mouth. We need you to hear what it is saying. A world full of billions of people who understand their eco-distress as super-fuel, and who have allowed their feelings to meaningfully reshape the deepest parts of themselves, is health by another name. It's also the most environmentally connected, socially just, and fruitful world we could wish to build from here.

PART ONE

FEEL IT ALL

1

THE PSYCHOTERRATIC STATE

I used to think the top environmental problems were biodiversity loss, ecosystem collapse and climate change . . . I thought that with 30 years of good science we could address those problems. I was wrong. The top environmental problems are selfishness, greed and apathy . . . and to deal with those we need a spiritual and cultural transformation. And we scientists don't know how to do that.

—GUS SPETH, environmental lawyer and scientist

On a small farm in the rural Australian hamlet of Goongerah, nineteen-year-old India MacDonell clutched her GoPro camera and trotted towards her father. He stood between some burning trees and the house he'd raised her in, wearing a hard hat and a personal protective suit. She tried to make her voice audible over the roaring wind and flames, and yelled out to him, "Shall I get the other pump going?"

"What?" her dad yelled back.

"What can I do?" she yelled louder.

"You can help with this hose," he replied, gesturing at one lying by her feet.

India picked up the hose and aimed it indiscriminately at the wall of embers in front of her, which rained down like a stream of fluorescent orange confetti. The air was thick with smoke and she retched, over and over, underneath her mask. "I can't breathe!" she yelled, but there was no time to go inside and get a better mask. The wind was bending the trees and dragging their branches, each one bright like molten lava, towards the house. Just a few feet more until they touched its external wall. The gusts were picking up speed.

India sprayed the branches haphazardly, let out a few uncomfortable squirms, and in a panicky voice said to herself, "I don't know what to do." It was the first time she'd ever used a fire hose. A moment later, the water pressure in her hose dropped dramatically; what had been a gush was now a weak trickle. She looked down at the hose, which was snaking on the ground. Two holes had appeared in its side where there used to be none, and water was spewing out of them. "Daaad! The hose is burnt!" she screamed at the top of her lungs.

The MacDonells' battle to save their home—which they miraculously succeeded in doing—lasted five hours, though viewing just a few snippets of it on YouTube provides enough material to haunt you for days. I was on the other side of the Earth while this was all happening, but my friend Catherine, who lives in Sydney, was also affected by Australia's infamous Black Summer, in which more than 12 million acres burned.

Catherine and I sent messages back and forth on the popular texting service WhatsApp as 2019 turned into 2020 and the flames did not subside. What I remember most from that time is the mixed feeling of despair and "survivor guilt" from knowing my friend was experiencing toxic smoke from record-breaking bushfires far away while I was in a safe location. My sense of helplessness made it hard to know what to say. The Bureau of Linguistical Reality has not yet coined a word for it, so I'd like to do so now: "whatsappathy." An excerpt from our thread:

NOVEMBER 20, 2019

C: It's scary :(omg so many dead animals and important populations of koalas burnt, and so many less cute animals
Super low air quality in Sydney like real foggy
It's ashy but smells like honey
People are wearing air masks and asthmatic kids not going to school [skull emoji]

Me: That's so crazy
Wow so sad :(

NOVEMBER 21, 2019

C: [Sends picture of glowing red sun in the sky]
Creepy red sun and grey sky over Sydney yesterday

Me: I've never seen a sun like that
It's strangely seductive
Quite beautiful
But obviously not OK!!
How are you holding up?

C: [Sends picture with clear view of a brown sky]
The sky looks like mud
Yeah, the sun was mesmerizing felt like the view from another planet, it was so big too!
Feels like a full-on climate emergency
I feel like "oh this is what is supposed to happen"
People are talking a lot about climate . . . a lot

Part despair but a lot of people changing their banks and stuff
Feels a bit desperate to try to do something
That was the sky in the middle of the day

DECEMBER 1, 2019

C: Fires still raging air smells like strong campfire and burns the nostrils
Had a friend that had to get emergency evacuated yesterday!
It's never ending!!

Me: Are you getting sick at all?

JANUARY 3, 2020

C: Happy new year!
We're still on fire
An old woman got off a plane in Canberra and died because the air was so bad
I keep writing "how do we fix it" then deleting it

Although bad environmental headlines regularly come in through our devices, which can deeply affect us, the digital scroll is also a flattening force. It renders our emotions into cartoons, limits the conventions of our language, and doesn't transmit the sensuous signatures of reality, like the smell of the air. That's not to say it doesn't create new possibilities for communication and empathy, but what an uncanny place to process historic disaster. Catherine tried to relay the gravity of the bushfires, and I tried to be supportive, but somehow the digital medium hindered us both and had the strange effect of minimizing the whole situation. Still, I understood the horror of the bushfires, and had listened to several interviews with survivors who described the unprecedented blaze as a literal hell. The whole continent was throbbing in pain.

Pain is a *natural* outcome of being told—and experiencing—that wildfires, hurricanes, and floods are becoming more ferocious due to the climate crisis, and that droughts are getting more serious and lasting longer. It is *reasonable* to get worried when the World Bank foresees that 140 million climate migrants will be fleeing ecological catastrophes and the knock-on effects of social strife within Latin America, sub-Saharan Africa, and south Asia by 2050, while other estimates put the number at over one billion. It is *normal* to get anxious about mass migrations and resource scarcity increasing the risk of violence and war. It is *appropriate* to grieve when the UN reports that humans are driving up to one million species to extinction, many within mere decades. It is *logical* to be horrified when a study shows that most trees alive today will be killed in massive die-offs within forty years if we don't dramatically change course. It is *understandable* to be scared when a different study finds that Arctic permafrost is thawing at a rate that was predicted to happen seventy years from now, that the Greenlandic ice sheet has already melted beyond a point of no return, and that unchecked climate change could collapse entire ecosystems as soon as 2030. Stress is a *suitable* reaction when scientists say that by 2070, one to three billion people will be living in hot zones outside the temperature niche that has allowed human civilization to thrive over the last six thousand years. It is *humane* to be gutted when you learn that air pollution caused the premature deaths of nearly half a million babies in their first month of life over a twelve-month period. It is *fitting* to freak out when the World Meteorological Organization issues a report on the global climate that states "time is fast running out for us to avert the worst impacts of climate disruption and protect our societies from the inevitable impacts to come." It is *sensible* to get spooked when a group of leading environmental researchers publish a paper that opens with the words, "The scale of the threats to the biosphere and all its lifeforms—including humanity—is in fact so great that it is difficult to grasp for even well-informed experts." It is *right* to be pissed off when you learn that we've emitted more carbon dioxide since the UN established its framework convention on climate change in 1992—that is, since we have been making an intergovernmental effort to reduce our emissions—than in all the millennia before then. And it is *decent* to rage once you understand how deeply the fossil fuel industry has manipulated the political system and misinformed the public, valuing money over our survival. There is nothing pathological about this pain. It is an unavoidable symptom of a very sick society.

It's natural to want to ward off this response, but it is precisely when we turn towards tough feelings and have the support to process them, that we unlock our capacity for strength and resolve to act. In *Earth Emotions,* environmental philosopher Glenn Albrecht writes that as damages ramp up due to economic systems that value profit over life, "the truth about fiscal irrationality and the insanity of deliberate climate warming will shock all the generations, and they will enter all forms of negative psychoterratic states. It will shake the foundations of current human identity to its core." What Albrecht calls *psychoterratic* are emotions related to perceived and felt states of the Earth. People are hurting because they can feel the Earth and its creatures hurting. I believe our species has entered this period, and we are in the early days of sensing all that it has awoken within us.

Numerous terms have emerged to describe the shaky psychological ground humanity now finds itself on. The most prevalent one is the aforementioned *eco-anxiety*, defined by the American Psychological Association as the "chronic fear of environmental doom." At this late stage in the climate crisis—a scientifically proven anthropogenic phenomenon that's been debated on baseless claims for decades—I would suggest eco-anxiety is merely a sign of attachment to the world. People often use the term interchangeably with *climate anxiety*, though it encapsulates wider registers of environmental concern. Eco-anxiety isn't listed as a medical condition in the *Diagnostic and Statistical Manual of Mental Disorders*, and many mental health professionals say it is important that it remains excluded. After all, the last thing we want is to pathologize this moral emotion, which stems from an accurate understanding of the severity of our planetary health crisis.

The climate crisis, the accompanying threat of ecological collapse, and the COVID-19 pandemic all mess with our sense of *ontological security*—the feeling of continuity in the order of our lives. For sociologist Anthony Giddens, ontological insecurity concerns a "person's fundamental sense of safety in the world and includes a basic trust of other people." Indeed, my eco-anxiety became significant a few years ago not only because the stability of the climate is clearly breaking down but because I largely lost trust in other people's abilities and determination to solve the problem. Our leaders have been either fully denying the crisis or promising paltry action that will only delay the inevitable, rather than addressing this emergency at the scale it demands.

Solastalgia, another term coined by Glenn Albrecht, describes "the homesickness you have when you are still at home," after undesirable environmental changes have occurred and you can no longer recognize home as the solace-giving place it once was. In light of recent temperature changes, Albrecht now pushes the turn of phrase even further, and writes, "Home is becoming more than unrecognizable: it is for many becoming increasingly hostile." *Ecological grief* is defined by scholars Ashlee Cunsolo and Neville Ellis as "the grief felt in relation to experienced or anticipated ecological losses, including the loss of species, ecosystems, and meaningful landscapes." Both those terms relate to *Anthropocene horror*, which literary critic Timothy Clark calls "a sense of horror about the changing environment globally, usually as mediated by news reports and expert predictions, giving a sense of threats that need not be anchored to any particular place, but which are both everywhere and anywhere."

Such emotional states can appear as a momentary flicker or a gang of feelings we cycle through. In many places throughout this book, I use "eco-distress", "eco-anxiety", or "climate anxiety" as interchangeable shorthand to refer to an assortment of challenging feelings a person can have after they've awoken to the planetary health crisis. Most often, though, in the day-to-day, we just call these feelings by their right names: fear, terror, anxiety, depression, despair, overwhelm, stress, worry, sadness, rage, grief, guilt, heartbreak, dread.

Global dread as defined by Glenn Albrecht is "the anticipation of an apocalyptic future state of the world that produces a mixture of terror and sadness in the sufferer for those who will exist in such a state." It's the dread that's been hanging over me for the last few years. It rains the heaviest when I speak with people much younger than me, like the student climate strikers, my nieces and nephews, or the unborn child I so deeply want to have. On some level, I see the dread everywhere I look. Congratulating friends who are newly pregnant has become a tightrope walk tinged with tragedy. The only way to fully revel in the joy is to consciously hold myself back from thinking about what the world might be like when that infant is my age.

When I speak with elders, and hear some of them glibly joke that they're looking forward to checking out of here before the shit really hits the fan, I feel enraged and abandoned. And when I take out the trash, I imagine every carbon-emitting step of where it might travel, as parts of it are separated out, put on trucks, trains, or boats, and whisked away to landfills and processing plants, God knows where (or actually, we do know where: typically, close to neighbourhoods where poor people of colour reside).

Casual conversations about climate disruption can become fraught with disagreement and resentment, and reveal a profound dearth of *emotional intelligence*—the ability to identify and manage one's emotions and acknowledge the emotions of others. As psychoterratic emotions flare, we must carefully consider the ways we get into these conversations and sustain them, with compassion and respect for people's lived experiences and multiple—often differing—truths. Our lack of emotional intelligence—especially evident on Twitter, where people scream at each other over the hot take of the day—is the foundation upon which sits the unprecedented polarization we've seen in recent years. Emotions are the gas you can't smell or see, yet it fuels the public engagement that drives social media and reaps profit for Big Tech. In the political arena, this missing intelligence is pulling at the threads of our democracies and families. And our lack of wisdom about our own feelings and those of others is just as damaging when we talk about what is happening to the Earth and all the lives that are at risk.

For a long time, those living in close connection with the land, such as Indigenous peoples and farmers, have been voicing despair as their identities and ways of life are unravelled by human-caused ecological changes. As a result of a legacy of racist policies, Black Americans are 75 percent more likely than whites to reside in "fence-line" communities that lie next to polluting oil and gas wells and chemical plants. A 2019 survey of Americans by the Yale Program on Climate Change Communication found that "Hispanics/Latinos and African Americans are more likely to be alarmed or concerned about climate change than Whites." It also found that they're more likely to get involved in environmental campaigns due to their disproportionate vulnerability. That's why, in the US, people of colour have been fighting for environmental justice for over a hundred years yet have not often seen it delivered, which understandably breeds rage, grief, and fatigue. Professionals who work to improve our collective predicament, from scientists to activists to environmental writers, have also been reporting for decades on the emotional toll of their work.

So this heaviness about the state of the environment is nothing new; what is new is the popularity of labels for ecologically linked distress and how much more frequently they are used.

In 2019, *Grist* magazine called climate anxiety the "biggest pop-culture trend" of the year. That same year, Oxford Languages reported a 4,290 percent increase in use of the term "eco-anxiety" compared with the year before and named "climate emergency" the word of the year. In 2020, a national poll carried out in the UK found that 70 percent of 18-to-24-year-olds were more worried about the climate crisis than they had been just one year prior. Meanwhile, a different survey conducted in the US revealed that 71 percent of millennials said that climate change had negatively affected their mental health. In the same year, academics created the first measurement scale to assess how climate anxiety manifests in people's lives.

Unsurprisingly, climate anxiety affects young people the most. They did not create this dangerous reality, but they have inherited it along with the duty to clean it up, and are often made aware of this before they've had the chance to figure out important aspects of their identity. Even if their present-day reality is comfortable, thoughts of a terrifying future can be disruptive. In 2020, more than half of eighty-two child and adolescent psychiatrists who were surveyed across England said that their young patients experienced this form of distress. The BBC surveyed two thousand 8-to-16-year-olds in the UK about climate change that year. Most children responded with worry about how the climate crisis would affect them as they get older, one in five had had a nightmare about it, and two in five said they did not trust adults to tackle the problem. The same proportion felt that their views as young people were being completely ignored. A survey conducted the same year by Våra barns klimat (Our kids' climate) found that six out of ten Swedish youth aged twelve to eighteen felt concerned about the climate and the future of the environment, only half felt hopeful about humans' ability to fix it, and that concern increased with age.

In 2021, my colleagues and I conducted a survey that looked at climate anxiety in 10,000 children and young people (aged sixteen to twenty-five) in ten countries around the world: Nigeria, Philippines, India, Brazil, Portugal, Australia, USA, France, Finland, UK. Of the global respondents, 45 percent said that their feelings about climate change negatively affect their daily life and functioning (this could include eating, concentrating, work, school, sleeping, spending time in nature, playing, having fun, or relationships). Over half said they think that humanity is doomed, that they won't have access to the same opportunities their parents had, and that the things they most value will be destroyed; 39 percent said they were hesitant to have children. Our study also showed that the psychological distress young people experience over the climate crisis isn't just about the degrading state of the environment. Rather, it is linked to perceptions of government betrayal and being lied to by leaders who are taking inadequate climate action while pretending otherwise.

Though the stats tend to show that this distress affects millennials and Gen Z more than their parents and grandparents, ecological dread is, of course, not limited to the young. Many older folks who are in touch with their environmental identity also feel wrecked by the unravelling they've had to witness over their lifetime. I write a newsletter about "staying sane in the climate crisis" called *Gen Dread*, and one of my readers expressed a sentiment I've heard from lots of older people when she emailed me to say: "I am 69, and the environmental changes that I have seen in the last 50 years of my life are unbearable to contemplate . . . There are many of us [the middle-aged and elderly] suffering in silence." They grieve for their grandchildren, grieve for the animals, and shake their heads in disgust at the state of the planet that's being passed on to the next generations.

Climate anxiety occurs across generational divides, appearing with particular intensity among the young, and it is marked especially clearly where racial and class privilege intersect. As journalist Abby Higgins writes in the *New Republic*, the "sudden proliferation of collective mourning" among citizens of the Global North "can obscure the stakes—not the stakes as they have suddenly become, but the stakes as they always were." Always were, for example, in Turkana, Kenya, where she reports that a lake which more than 300,000 people depend on is drying up. One might imagine that climate anxiety would be rampant in such a place; however, the term is not as prevalent in parts of the globe where the climate crisis is already most felt. One study found that the only people typing "climate anxiety" into their search bars were in Canada, the United Kingdom, America, and Australia. Its striking absence from Internet searches in the Global South (including the English-speaking parts) calls into question the suitability of a term like "climate anxiety" to describe people's feelings there, where myriad other social injustices, legacies of environmental violence, and poverty intensify the psychological distress people are experiencing.

But it would be a mistake to conclude from this finding that this term isn't applicable in the Global South, period. Although early discussions of climate and eco-anxiety have tended to focus on citizens of the Global North, which adds to the sense that they apply to a privileged segment of humanity, the breadth of focus is starting to change as more researchers become interested in the field. Several are now looking at how these concepts tracks in the Global South as well as how they land in communities of colour within the Global North.

Jennifer Uchendu is a young Nigerian climate activist who has studied eco-anxiety in London as well as her hometown of Lagos. She sees clear differences between how young Brits relate to eco-anxiety and how she and her friends back home do. Young people in the UK, she says, often feel guilty about being citizens of an industrialized nation that's making the problem worse and have a troubling sense that they can't do anything to fix it. Young Nigerians, on the other hand, often feel angry about the environmental injustices and climate impacts they're already experiencing, as well as fatigued by the issue. However, learning during her studies that "eco-anxiety" was a term brought Jennifer massive relief. It explained why she was burning out on her activism with the environmental organization she founded called SustyVibes, and why she was losing faith in her work even though passion had always previously driven her efforts. She told me, "Every time I talk about eco-anxiety on social media, young people can immediately relate to it. They know what I mean. I remember when I came back home and talked to our volunteers, they were like, 'Oh my God, it has a name, it's a thing, people have figured it out!' And that was exactly how I felt in the UK, realizing that this is actually valid. I think that's the word—*validating* these feelings and saying I'm not crazy, I'm not overthinking, I'm not taking it too far. This is a real problem."

It is in the United States, though, that the conversation about climate anxiety being an unbearably white phenomenon has really got off the ground. In *Scientific American*, professor of Environmental Studies at Humboldt State University Sarah Jaquette Ray writes, "Climate anxiety can operate like white fragility, sucking up all the oxygen in the room and devoting resources toward appeasing the dominant group." *White fragility* refers to the defensiveness and discomfort white people often express when confronted with the traumatic impacts of white supremacy on people of colour, which their privilege protects them from feeling the need to dismantle. In other words, it is what's on display if white people react from a bruised sense of self when faced with the anger of people of colour, rather than focusing on the real harms that racial inequality causes in their lives. It is damaging because it leads to white outbursts or white shutdowns, both of which prevent the kinds of dialogue and partnership that are required for racial justice to be achieved. Ray thinks that climate anxiety is like white fragility in that it can cause white people to turn inward instead of looking outward to understand who they are in this uneven and ongoing cataclysm.

Navel gazing isn't the best route for scrutinizing the role one plays in upholding larger systems that make people of colour disproportionately vulnerable. Ray makes the good point that racial inequality and the climate crisis need to be healed at the same time because they are inextricably intertwined. Hop Hopkins of the Sierra Club underlines this relationship well: "We will never survive the climate crisis without ending white supremacy . . . You can't have climate change without sacrifice zones, and you can't have sacrifice zones without disposable people, and you can't have disposable people without racism."

Turning inward can lead to a myopic focus on the self and an obsession with the protection of one's own, through things like borders, bunkers, or all-out brutality. Ray, who has also written a book on climate anxiety, told me, "I am genuinely worried about how anxiety about climate change—and any big feelings about the environment for that matter—can result in xenophobia, exclusion, and even violence." Ray cited the massacre of twenty-two people in El Paso in 2019 by a shooter who declared he was trying to stop the "Hispanic invasion of Texas," and whose manifesto read, "The decimation of the environment is creating a massive burden for future generations . . . If we can get rid of enough people, then our way of life can be more sustainable." Ray sounded the alarm on climate anxiety as a possible on-ramp for fragile white people to drive straight into the arms of eco-fascism. Indeed, the shooter who killed fifty-one people at two mosques in Christchurch, New Zealand, referred to himself as an eco-fascist in his manifesto, meaning he ascribed to an ideology that says the best way to deal with climate change is by vanquishing migrants. These are terrifying developments that do not bode well for how the far right might deal with rising temperatures and dwindling resources in many countries in the years to come.

Underserved groups must often deal with other pressing existential threats that mix in with their climate anxiety, while the most privileged and protected may focus their attention on just this one issue. Jade Sasser, a professor of gender and sexuality studies, has been interviewing young Americans of colour about the existential threats that inform their reproductive decisions. She tells me that the climate crisis is certainly on many of these young people's lists, but it usually appears below ongoing threats to their well-being such as racial violence and concerns about how they would be able to economically provide for a child. Climate anxiety affects different groups in different ways. At the same time, our survey of 10,000 young people around the world, mentioned earlier, demonstrated how prevalent the thoughts and feelings associated with climate anxiety are in a range of cultures and climate vulnerabilities across the Global North and South. This is nuanced stuff, but Ray's main point about the connection between big environmental feelings and white fragility remains: privileged anxiety about the climate, such as my own, must be harnessed and purposefully directed outward for justice-oriented results if it is going to be of help.

ANTHROPOGENIC CLIMATE CHANGE is commonly regarded as a consequence of human development, a phenomenon that results in rising temperatures due to increased greenhouse gases in the atmosphere emitted by industrial activity. But Indigenous and anti-colonial scholars have shown how the climate crisis is also an exponential amplification of the environmental harms inflicted upon Indigenous peoples for more than five hundred years. Global warming is not only a natural result of burning fossil fuels, it is an extension and outcome of colonization as well.

European colonization of the Americas is inextricably connected to the dispossession of Indigenous peoples of their land, and to the global expansion of fossil fuel use. For example, before the Industrial Revolution, southern China and northwest Europe were comparable in their material consumption, including the use of coal. European annexation of the Americas allowed the continent to massively transcend its physical boundaries and propel an extraordinary inflow of cheap raw materials from Indigenous lands for industrial processes, including exploitable coal reserves. People stolen from Africa, enslaved, and forced into labour for the creation of the "new world" were indispensable assets. The Industrial Revolution evolved from these expanding conditions for European productivity, as did the bedrock of unequal social relations between white, Black, and Indigenous peoples.

Settler colonialism profoundly disrupts the intimate relationships of Indigenous communities with plants, animals, water, and minerals. Alongside the intentional slaughter of Indigenous peoples to enforce settler control in places like my home country of Canada, there was and is the decimation and invasion of ecosystems and species that Indigenous peoples consider kin. This is the deep environmental impact of colonization—development for some, destruction for the rest—which to this day informs the emotional state of our world. We still see the effects of this history all around us, and we'll only be able to improve the situation if we clearly understand where these traumas—to the land, to species, to people—come from.

Though many Indigenous peoples died as a direct consequence of colonization, those who survived had to learn to adapt amidst new environmental and social pressures. This is why Kyle Whyte, an environmental justice researcher, activist, and member of the Potawatomi nation, notes that as the fear of climate chaos escalates amidst the public at large, along with concerns about how to cope, for many Indigenous peoples, this all feels like déjà vu. What we're dealing with is a cyclical repetition, but this time around, we're all affected. Many more people are now in harm's way, and disturbingly, governments are preventing progress towards more livable futures where people stand a chance of adapting.

I used to think about the mental health impacts of the planetary health crisis in binary terms, with a clear separation between those who lose their livelihoods from drought, their homes in wildfires, or their cultural traditions as a landscape erodes, and the rest on the sidelines, who feel anxious from reading about those events in the news. But as my research piled up, the porous nature of this imagined divide became clear. The stress of what's happening to the natural world is living inside people's bodies. Of course, it is important not to minimize the vast differences that exist in how climate trauma manifests. Disasters that affect people directly exacerbate mental health disorders; droughts, hurricanes, heatwaves, floods, and wildfires have been shown, time and again, to spike post-traumatic stress disorder, anxiety, depression, suicidality, substance abuse, and other mental problems. But eco-distress is not some entirely disembodied experience. It can impair functioning even if one is far from the front lines—causing physical symptoms such as sleep disturbance and panic attacks. Climate-aware psychiatrist and historian Gary Belkin put it to me this way: "The whole frickin' world is a disaster zone, a crime scene, there are no sidelines." Though some of us are more privileged and protected, and will continue to be because it is not a level playing field, we are all on the field.

Before the COVID-19 pandemic, for the layperson, mental health was largely seen as a problem that individuals must grapple with inside their own personal orbit. If a person was feeling suicidal or depressed, their pain might be attributed to whatever happened to them in their life, or whatever biology they were unlucky enough to inherit. Poor mental health hasn't been principally understood as the consequence of biased societal expectations—how we grow the economy, fund public health services, or fail to take care of one another. But the pandemic has made us notice something new.

People everywhere—and their governments—are starting to see mental health as a public health issue in a way they hadn't before. As populations were forced to stay at home to avoid the virus in the streets, millions around the world experienced pervasive anxiety about getting infected, losing family members, and being separated from all that had once felt normal. For some, it became hard to sleep, while for others it was unbearable to be at home 24/7 with the demands of child care and home-schooling; and for a great many, the goodbyes they made to loved ones on ventilators through smartphones and virtual funerals massively complicated their grief. Front-line workers who risked their lives every day died in disproportionate numbers, which resulted in more deaths among communities of colour.

As with climate change, young people have been the most susceptible to despair during the pandemic. After four months of the new world order under lockdown, a survey conducted by the Centers for Disease Control and Prevention found that a quarter of 18-to-24-year-olds in the United States had seriously considered suicide. A UNICEF survey of eight thousand young people in Latin America and the Caribbean found more than a quarter experienced anxiety related to COVID-19, while more than a quarter of seventy thousand French students surveyed experienced depression, and 10 percent of them had suicidal thoughts during the pandemic's early months. This mounting evidence led the *New York Times* to call the situation a "mental health pandemic" that should be treated as seriously as containing the virus. The pandemic has rendered clear how dependent our mental health and well-being are on the strength of our shared social fabric.

And yet the pandemic is a mere dress rehearsal for what scientists anticipate the climate crisis will do to our well-being. We are told to expect even more daunting existential challenges to come our way in the next decades, which will affect the resilience of our communities and our own interior worlds. Inversely, how grounded we are as individuals and how prepared our communities are will determine how we respond to these challenges. We cannot afford to be caught unawares as we were with the COVID-19 pandemic, when at its inception mental health was not a political priority and was chronically underfunded in most countries. We've seen this negligence exacerbate pre-existing inequalities, and disproportionately threaten vulnerable communities. The climate emergency does the same, and although its harm is already in motion, a sober look at the coronavirus pandemic presents us with an opportunity to better mitigate the mental health impacts of a warming world.

When we recognize the climate crisis for what it is—a collective trauma—we can begin to invest in our own mental health and well-being in relation to it. And the sooner we begin this process, the better, before people are even more drained from the stressful news and directly felt disasters of the day. We are *all* at risk of becoming traumatized by what is happening, and would be wise to recognize that anything that helps others is a way of also helping oneself.

The accumulating effects of colonialism and various manifestations of eco-anxiety have awoken the psychoterratic state. We are presented with one overarching question: Are we going to let our feelings overrun and deplete us, or are we going to use our feelings to overrun the systems that are making us so unwell? If we don't recognize our eco-distress, and learn to purposefully and justly transform it, we'll miss the opportunity we have to use all this restless energy for a deeper purpose: to bring forth the future we owe each other by promoting solutions that have co-benefits for all people and the planet. Many now appreciate, at least intellectually, that the planetary health crisis is under way and that it is a real threat to their health and well-being, but they aren't letting it settle into their souls. They need to. The distress this realization kicks up has vital tools to give us, and it is our job to listen.

Anxiety and grief about the world can be a positive thing. Dr. Martin Luther King Jr. often discussed the merits of being maladjusted to a society that values segregation, discrimination, militarism, and inequality. "The salvation of our world lies in the hands of the maladjusted," King said in 1955. In a speech ten years later, he said, "Through such maladjustment, we will be able to emerge from the bleak and desolate midnight of man's inhumanity to the bright and glittering daybreak of freedom and justice." Being maladjusted does not mean you are sick when the maladjustment is in response to a sick society. It serves an important function by pointing out what is deeply wrong, and gets you to try to change your situation. Eco-anxiety, solastalgia, ecological grief, and other difficult eco-emotions are the fastest-growing symptoms of maladjustment of our time. We should be proud and relieved to feel these emotions; they are a sign of our humanity.

Our maladjustment may also be the only true resource we have left in our rapidly narrowing time frame to protect much of what can still be saved and plant the seeds of renewal. As a group of climate change and mental health researchers wrote in the *Lancet*, one of the world's most respected medical journals, "Recognizing that emotions are often what leads people to act, it is possible that feelings of ecological anxiety and grief, although uncomfortable, are in fact the crucible through which humanity must pass to harness the energy and conviction that are needed for the lifesaving changes now required."

As I see it, the task at hand is to create new norms around how we handle and harness psychoterratic emotions for the sake of our health, well-being, and collective survival. How can you stay present and engaged with an environmental reality that is becoming increasingly difficult to bear, even when perceived through digital mediums that both spike and limit emotional response? What kind of support do you require in order to be able to consider the Earth, other people, and other species in your actions, when it is so much easier to look away? These questions aren't probing for technological, political, or scientific answers, as much as they are trying to draw out life-protecting dimensions of your psychology, culture, and spirit. This work is existential, and it is all of ours to do.

KEY TAKEAWAYS

» Eco-distress is a sign of connection to the world; it is a normal reaction to the injustices being inflicted upon the planet and its living creatures.

» Climate anxiety is especially visible where racial and class privileges intersect, and the most privileged have a responsibility to harness their feelings for positive change.

» A long history of colonialism is inextricably linked to climate catastrophe, the destruction of Indigenous cultures, and planetary harm.

» Eco-distress is an embodied phenomenon that affects everyone, everywhere, though it may be called by other names.

» The COVID-19 pandemic has put mental health on the map as a public health issue; this awareness can be used by policy-makers to get ahead of some of the (vastly more abundant) mental health problems that the climate crisis will cause over time.

» Being maladjusted to a sick society can be a source of strength for taking on the life-protecting actions that are needed now.

2

THE POWER OF DENIAL

Humankind cannot bear very much reality.
—T.S. ELIOT

When we are faced by difficult truths
we are susceptible to lies (and liars).
—PAUL HOGGETT, climate-aware psychotherapist

In the backyard of a small bungalow in the East York neighbourhood of Toronto in October 1962, a man, woman, and baby wait patiently to be killed. The infant, blissfully ignorant of everything except the grass under his palms and knees, is my eldest brother, Michael. The man is my father, Joe. He is sitting in a lawn chair next to his first wife, Gail, and a bottle of whisky. The media have been saying that nuclear confrontation is inevitable.

Newspapers have been printing spy plane photos of military bases in Cuba, identifying vague shapes as missile launching sites. Authorities have encouraged people to dig survival bunkers in their own backyards. My dad has read the diagrams on how to do it. They detail what amount of radiation would be required to penetrate which size of walls, what kinds of foodstuffs he could store there, and so on. But he can't afford a bunker—he's only twenty-three. Instead, he replays in his head the radio broadcasts that said a Russian military ship was sighted off the coast of Florida, hauling a nuclear missile. Word is that the Russians have already surreptitiously installed nuclear weapons in Cuba, and they will be released tonight. The targets are said to include New York and Washington, either of which will wipe out Toronto according to the concentric circles of radiation. My dad believes they are doomed to die a fiery, blistering death this evening. He touches his wife's and son's smooth skin one last time, looks up at the sky, and waits.

"HOW DOES IT feel to throw your own end-of-the-world party?" I asked my dad fifty-seven years later.

"It was numbing, because we were totally exposed," he replied. "We had no recourse, any way of saving ourselves. We believed that the release of nuclear weapons was imminent and that we would be fried by radiation and that it would be a very unpleasant death."

"Yes, but what does that numbing do to you?"

"The sense of tragedy was so completely overwhelming that we couldn't even talk about it. It was just so real, so palpable, that we were literally dumbfounded. It was too immediate and too overwhelming to discuss. I mean, it was unspeakable horror, so literally there was nothing to talk about."

Having heard this story my whole life, about how the Cuban Missile Crisis touched my family, I've always found it hard to understand why they wouldn't have talked to each other at length, performed some kind of ritual, taken stock of life, or at least stared into each other's eyes until what they perceived to be the last moment. Then again, what do I know? I've never believed I was going to be burnt to a crisp within the hour.

"It was just too much. I mean, I blanked out," my dad continued. "I couldn't deal with it relative to what the environment might be like after. The injustice of it all was overwhelming. Gail and I, as I recall—to the extent that I recall, and maybe I don't recall properly because of the fact that it was overwhelming—I think we didn't talk about it, period. We just sat there like fucking dummies."

Sitting ducks.

Until my awakening in 2017, when Sebastian and I began to talk seriously about having a child, I never considered myself a climate activist. After all, I have never protested a pipeline or been arrested for direct action. I was not at Standing Rock or any UN climate meeting. But I was environmentally conscious and I was doing my part, I thought. I prided myself on not owning a car, biking everywhere, and, only if I had to go really far, using public transit; it was my way of chipping in, and it felt conveniently noble considering it was all I could afford in my student days. When I was a teenager, I saw a friend rinse out an empty milk bag—that curious liquid holder only Canadians recognize—and reuse it to store leftovers in. That made her the patron saint of recycling in my eyes, so I started doing that too.

I dumpster-dived for day-old baked goods and unsold fruit crates on several occasions in my early twenties when I was living in Montreal, motivated by a desire to fit in with my anarchist-adjacent friends more than an ethic of food sustainability. I remember feeling relieved a handful of years later when a different friend told me she was only going to buy vintage clothes from now on, for environmental reasons. Phew, another environmental box I could tick. I already bought most of my clothing used, since fashion from the seventies is my favourite. Then I went vegetarian, sort of. And yes, as I grew older I was flying way too much, and no, I wasn't offsetting every ticket, but I was at a handful of climate marches over the years. I was doing my part, okay?

I was a sitting duck too.

One could argue that we've been in that backyard looking up at the sky for roughly forty years, waiting for this other apocalypse to get us. We haven't built a bunker, created any goodbye rituals, or professed our love to each other or to life as a whole. Given our mass inaction on climate, and the fact that we are now facing hysterical levels of loss, suffering, and pain, many commentators like to say that there is something in our nature that makes us not care about climate breakdown.

Our cognitive biases have evolved to save us from imminent danger. Facing the possibility of attack by a wild beast at any moment, our ancestors learned to overvalue immediate threats at the cost of discounting more far-off and abstract perils. This adaptation is still with us, and shows up when we panic over dramatic events like the exponential spread of a virus around the globe but don't take the steps to defend ourselves from steadily worsening climate disruption that promises to kill many more humans. We tend to stick with whatever we've invested time and resources into, like the fossil fuel system we've built, and ignore clear evidence that it's harming the planet. For years, reliance on oil and coal has persisted despite the obvious environmental benefit and increasing affordability of transitioning to clean energy. Many of us are steadfast in our reluctance to overhaul the system because we perceive it will lower our standard of living in the here and now for the goal of creating a hypothetical survivable future that we can't yet feel. And these fibres of our being are now braided together into a tight noose that is strangling us.

The psychic numbing my dad experienced the night he was sure he'd die is the same response animals exhibit when they play dead in the face of a threat. The psychiatrist Robert Jay Lifton, who studied the psychological causes and effects of extremism and war, has described this phenomenon. When one faces an overwhelming threat that is inescapable, and when one also can't muster what's needed to resist, numbness takes over. And though the immediacy and sensationalism of the nuclear and climate threats differ, they have both led to numbing. "The widespread numbing created by nuclear and climate imagery of extinction can be understood as playing dead on the part of the majority of people on earth," Lifton writes.

Why might this be? They are, after all, very different threats. With nuclear, only a handful of people ever have anything to do with setting off a warhead, whereas climate change is built into our world, made worse by each item of clothing we buy, each combustion engine car we drive, each vacation we take. I have never known a day on Earth when my actions didn't fuel the fire, literally, and the fact that we can see how we contribute to this problem that makes us feel unsafe is crazy-making in itself. With nuclear, we have a binary situation: detonate or don't. With climate, we have the fabric of the world, and pulling threads is not the same as disarmament.

Martin Fleck is the director of the Nuclear Weapons Abolition Program at Physicians for Social Responsibility, where he has been working since 1981. Around that time he remembers saying, "I have to do something about the nuclear threat, because if I have kids and this stuff goes to hell one day, I will never be able to face them." Yet he doesn't remember many people in the nuclear heyday talking about *not* having kids, or at least not as much as they do today with respect to the climate. "Perhaps the difference now is that climate change is right there for everybody to see. It is visible, and happening, something we read about in the news every single day. Whereas with nuclear, we mostly had to imagine the threat to get scared," he told me.

Though nuclear is still very much a real threat today, young people don't typically tend to get energized by it. One day, a millennial woman Fleck used to work with helped him understand why. She said that his people, the nuclear folks, with their constant message of "we've got to prevent the apocalypse," talk right past millennials and anyone younger. Because of the climate crisis, today's young people think the apocalypse is already happening, that it is currently under way, and there is no such thing as preventing it. The "done deal" sentiment of it all easily leads to psychic numbing.

The danger in comparing these threats is that we are not equally helpless in the face of them. If a foreign leader suddenly drops a bomb on you, there's really nothing you can do. But when equipped with decades of measurements and models and infinite choices for systems change, as we have with the climate, the helplessness we feel is a sham. Every decision counts, and from every moment forward, each choice has the potential to contribute to the prevention of more suffering. We've learned to maintain our helplessness.

Timothy Morton, sometimes referred to as the "philosopher prophet of the Anthropocene," says our inability to deal with what's happening has to do with how climate disruption defies what it means to be a thing. It is so vast and all-encompassing that it transcends the normal valence of thingness altogether, making it what he calls a *hyperobject*. A hyperobject is distributed across space and time to such an extent that our minds cannot fully comprehend it. It is both everywhere and between everything, from fish in the sea to infection rates of vector-borne diseases to struggles of generations yet to be born to national defence spending to melting glaciers. It is impossible to think about as a contained phenomenon. By not seeing where the hyperobject starts or ends, we don't know how or where to intervene to cause meaningful change. At its simplest, this philosophy is a more interesting and artful way of articulating the cognitive limitations that hold us back, which psychology and psychoanalysis have long held to be the keys to understanding.

Human beings are as irrational as we are rational, and we are rarely transparent, even to ourselves. Robert Jay Lifton, with his studies of Vietnam War veterans, Chinese thought reform, and Nazi Germany, enhanced our understanding of how humans are capable of accepting evil acts. His work was built upon the understanding that there is no such thing as a single unitary self, the rational part that sees all, knows all, and operates accordingly. We are made up of multiple differing selves, many of which are irrational and often in conflict with each other. The best way to grasp why we are not responding to the overpowering threat of climate change is to look at those contradictory selves. As climate-aware psychotherapist Paul Hoggett writes, this approach is focused on "how we avoid, deny, embrace or accept it, dream about it, get depressed, terrified or guilty about it, feel in two minds about it, think it is a sign of the Second Coming or Homo sapiens' final and deserved comeuppance, ruminate about it, wake up at night because of it, can't get our head around it, feel that 'really climate change is such a drag,' know that it is something one should worry about (even though one doesn't), and so on and so on."

Looked at in this way, climate denial is not only reserved for the fossil fuel companies, lobbyists, and corrupt politicians that are still spreading the damage; the truth carries a heavy psychic toll for all of us. Unless carbon emissions are drastically and quickly reduced, terrifying outcomes are inevitable, and already, no matter what we do, many further losses await us. Most of us don't deeply engage with this reality, not because we don't care, but because it is painful. If we don't have a safe place to express and explore the way it makes us feel, we become vulnerable to *environmental melancholia*, a term coined by environmental psychologist Renee Lertzman. It describes the feeling of conflictedness that can occur when confronting environmental problems. On the one hand, we aspire to contribute to solutions and help regenerate the Earth. On the other, we're completely overwhelmed by the eco-crisis, and many of us can see how our actions exacerbate the problem. Our unprocessed anxiety and ambivalence then shut us down, underpinning what she calls the "myth of apathy." It isn't that we aren't concerned, Lertzman argues; we're just so deeply caught in the double bind that we become immobilized.

If you work in the airline industry, for example, or factory farming, or simply drive a car with a combustion engine, embracing climate science necessarily involves dealing with your own complicity, and this elicits a taxing dilemma. It is much easier to defend yourself from inevitable discomfort by detaching from reality in your lived experience. That's why denial is a much more subtle and inclusive process, and one that most people unwittingly participate in. It comes in a few different forms, including outright denial, negation, and disavowal.

Outright climate denial is often ideological and vested in corporate, political, or other interests. It is also flexible, capable of existing at the level of institutions or individuals. Groundbreaking work by science historians and investigative journalists has shown that fossil fuel companies knew the real science decades ago but chose to fund climate denial to distract attention from their bad but profitable behaviour. Exxon, for example, funded its own climate modelling group in the late seventies and early eighties, and even confirmed the dangers of the greenhouse effect.

An internal Exxon memo from 1981 reads "it is distinctly possible that the corporate planning scenario will later produce effects which will indeed be catastrophic at least for a substantial fraction of the Earth's population." Not only did they know the science, they knew it was going to get, in their own words, *catastrophic*, if emissions were not curbed. They also knew that the public understanding of this fact could kill their business. So they did what any amoral corporate entity would do, and started funding denial and disinformation to protect their returns.

Science historians Naomi Oreskes and Erik M. Conway unveiled the nefarious link between climate denial and Big Tobacco in their book *Merchants of Doubt*. Big Tobacco used a smattering of wily tactics to distract and confuse the public on the deadliness of cigarettes by funding studies about all the other causes of lung cancer, and the oil industry learned from them the art of being an illusionist. In some cases, the exact same men who'd sown doubt about tobacco were hired to do the same about fossil fuels. By planting fake experts, cherry-picking data, exaggerating uncertainties, and feeding conspiracy theories, the oil industry directly copied Big Tobacco's playbook for interfering with regulation and public distrust.

In her investigative podcast *Drilled*, climate reporter Amy Westervelt shows that the manipulation started long before fossil fuel companies learned to spin the science. As early as 1914, in an attempt to manage controversy around John D. Rockefeller's Standard Oil Company, the oil industry was funding the evolution of agenda-driven public relations strategies including disinformation campaigns, fake news, message control, and PR crisis management. While fossil fuel companies would later come to spend a lot of money on denying the science, they spent magnitudes more on PR that nudged people to associate oil with progress, American exceptionalism, the good life, liberty, and happiness. A string of military intelligence men became the "Mad Men" of the American oil and gas industry, grooming executives and companies for public acceptability. Over time, they taught us all to use the language that benefits corporate interests, which is not a uniquely American phenomenon. For instance, in 2005, British Petroleum ran a media campaign that cost over $100 million and promoted the idea of a "personal carbon footprint," encouraging consumers to lower theirs, effectively placing the blame on the individual rather than the corporations that profit from emissions. Many environmentalists have unwittingly used this same terminology in their own campaigns.

And now there's an eruption of denial at the same time as a mass awakening. Because denial is based in fear, as the climate crisis worsens and more student strikers and rebellious activists raise hell (in the streets and virtually), deniers, ever more incapable of facing what threatens them, are clamping down doubly hard. The abusive conservative backlash to Greta Thunberg—the Swedish teen activist who catalyzed the global student strikes—has included Murdoch-owned papers ridiculing her Asperger's, cyberbullying, and right-wing websites preying on her parents. All of these are forms of abject denial. And as planetary health gets worse, denial will get worse too.

Another psychological defence—*negation*—comes from our ability to fully push away what is real. If you've been repressing a fact that fights its way into your consciousness, but the consequences of that fact are denied, then you've negated it. When it comes to negating the climate, the developers now moving at breakneck pace to build homes in the flood-risk zones of South Miami and Miami Beach are ignoring the inevitably short-term lifespan of those homes—and the potential danger to their occupants. Rising sea level is thought to have played a role in the collapse of the Champlain Towers South, a seaside condo that made headlines in 2021 for strewing residents and concrete blocks across Miami Beach.

But the most prevalent defence against climate breakdown is *disavowal*, a kind of soft denial. Disavowal is like having one eye open and one eye closed at the same time. Consider this: We believe the science, understand the risks, and are concerned about systems collapse, and then in equal measure we play down the threats so that we can continue to live out our lives according to our desires. We read that our unchecked warming means we are fast approaching tipping points that, once surpassed, will set off a cascade of impacts across biophysical systems *that cannot be reversed*. Then simultaneously, we do not believe that things so terrible will ever come to pass. Sound familiar? Many of us live out our daily lives in this conflicted place. I certainly do sometimes, with the flights I've taken to attend environmental conferences where I talk about my work. Different parts of us split off from each other, each loyal to contradicting values and aspirations. If done consciously and with measure, it can be beneficial at times to engage in a bit of short-term suppression in order to get through the damn day. But when we don't clearly acknowledge that we're doing it, we get stuck, and are unable to make the meaningful changes that we must.

Defences like denial are, after all, psychological adaptations for survival. Throughout history, humans have been able, with a dash of wishful thinking and self-delusion, to keep going when reality feels intolerable. The problem is, these various kinds of denial can manifest into what psychoanalysts call ***perverse states of mind***. What's perverse is how different parts of the self can form lying relationships with each other to protect us from anxiety and pain, regardless of the cost. These lies then sustain a form of "psychic retreat" from reality. The Oxford dictionary's definition of "perverse" is "showing a deliberate and obstinate desire to behave in a way that is unreasonable or unacceptable, often in spite of the consequences," which maps beautifully onto how we've dealt with our climate predicament.

This self-protective measure has dire consequences that disproportionately affect people who don't have the luxury of engaging in the same kind of thinking, because a life-threatening scenario simply winds up at their doorstep. It's obvious that our governments, companies, institutions, and many individuals—particularly those with privilege—filter life through perverse states of mind. Artfully engaging with the truth, refusing to accept responsibility, and continuing to enact a carbon-intensive world order, corporate plan, or lifestyle are all examples of this self-protective thinking, which is a form of exceptionalism.

Exxon advertises that they're on top of the climate problem so we shouldn't worry, and our leaders attend intergovernmental climate meetings where they perform national success based on emissions *targets* rather than actual *accomplishments*. If you're cynical, Paul

Hoggett points out, you could see something like the Paris Agreement as a kind of perverse governance. I don't, because I know how hard such accords were fought for, but the non-binding "we won't if you won't" pinky swears fuel the fantasy that things are on track rather than completely off the rails. The deals made at these distinguished gatherings disavow a much darker reality—that we're nowhere near meeting those targets and are failing to keep our promises.

I wanted to understand this perversity better, so I called up climate-aware psychoanalyst Dennis Haseley. He helped me see that, once you know what to look for, the perversion is everywhere. Say you're watching a football game on TV and then a commercial comes on. What's on the screen? Quite often, a lot of big, gas-guzzling trucks. These commercials become a way of warding off the force of the climate crisis, denying the reality that it's harmful to invest in a future filled with combustion engine vehicles. "People are saying, 'Yeah, we can buy trucks, I love trucks, I want the one with 460 horsepower!' and I'm sitting there thinking, you know, what the fuck are you doing? Don't you see what's happening?" Haseley said. The truck is the desirable object that helps us pretend.

The incessant truck commercials are like being tied to a train track and knowing that train is coming, but it is still two hours away and you respond by saying, "Let's have a party and pretend that this is not happening! Could someone please pour some tequila down my throat?" As Haseley explained, "When something is coming at you that raises anxiety, the hope would be that there would be some resilience that would help you try to untie yourself, not pretend yourself away from the pain until you're killed."

We have developed peculiar neuroses about this global threat. It can feel more authentic to connect with a sense of helplessness than an ability to make a difference. What psychologists call *self-efficacy* is the belief that we can create the outcomes we want by taking certain actions. It lives in us as confidence that we can achieve our goals. But if we've tried hard to achieve certain goals in the past and failed, this reduces our self-efficacy, and we experience ***learned helplessness***. If we have a high degree of learned helplessness because we've failed many times before, we might think we could never make a meaningful difference in addressing a huge challenge like the climate crisis. What's weird about the learned helplessness we see in society regarding this threat is that we've never really failed because we've never really tried. Our narrative of powerlessness is holding us back more than our actual capacities.

It turns out that the sour cocktail of denial and disavowal with which many respond to the climate crisis is not just a personal defence but also the natural outcome of a *bubble*—which psychoanalyst Sally Weintrobe defines as a world of make-believe, constructed by powerful people whose perverse aim is to diminish the moral unease, rage, grief, and anxiety that the citizenry would feel if the conditions of reality were made clear. Bubbles promote the unsustainable way as the only attainable and reasonable way to live. The life force at the heart of the bubble is a "culture of uncare," she notes, which maintains and promulgates the social acceptance of selfish impulses and short-sightedness. Neoliberal policies have been the greatest boon to the culture of uncare, as they promote free trade and individualism while denying the importance of social welfare. The truck commercials, overstocked (and waste-producing) grocery stores, and environmentally damaging fast fashion industry are all examples of the culture of uncare. They reinforce an attitude that invites and permits people to disconnect from parts of themselves that take responsibility in life.

Powerful people construct bubbles for profit using advertising, propaganda, and the reinforcement of certain social norms to make sure that when the bubble pops and fraud is revealed, it is not they who suffer. Weintrobe points out that when the financial bubble of 2008 popped, bankers made money while six million Americans lost their homes and austerity measures upended people's lives in many parts of the world for years. The political establishment's and fossil fuel industry's long-standing knowledge of the dangers of the greenhouse effect is no different. In order to profit, they had to create a climate bubble. "The climate bubble is far larger, far more consequential, and far more damaging than any bubble so far in human history and the fraud involves extracting all for now and leaving life bereft of a future," Weintrobe said at a climate psychology conference convened by the Freud Museum.

A striking feature of the climate bubble is that it has generally denied permission to talk about the climate crisis in emotional terms, particularly in highly influential places. Many scientists are extremely careful not to come off as a climate advocate or activist because they know they'd be punished for it, by having their research labelled as non-objective. This neutrality is achieved most easily by not speaking in energetically charged terms about the terrifying implications of climate disruption. Camille Parmesan, who has served as the lead author for the UN's Intergovernmental Panel on Climate Change, explained to me, "The scientists that I know of who become activists, their research is still very unbiased, very rigorous. But I've met enough politicians or staffers who say, 'Well, we can't listen to what so-and-so says because everyone knows that they are on the side of climate change disaster.' So, hearing that my whole research life, I've been really careful to not be viewed as an advocate by the people I'm hoping my research will influence."

It is completely understandable that a scientist wants their research to make its way into policy so it will effect some positive change in the world, and therefore they do what they must to protect it. But the dehumanization of scientists in the halls of power, which denies that they can both be real people with emotions about what they find out and also be working from impartial evidence, reflects a wider taboo inside the climate bubble, which says that some truths should not be named. How are average people with relatively limited scientific knowledge supposed to confidently support the grave implications of this research when scientists can't speak their full truth?

I first woke up to the perverse potential of environmental media at a film festival in Toronto. It was 2011, and I was watching filmmakers pitch their documentaries to funders. We saw a moving trailer for a documentary about orangutans in Borneo that were being wiped out by deforestation. It was appalling, and the filmmakers captured that drama. When it came time for potential funders to respond, the director of one of Canada's largest environmental foundations spoke first, and said that they would not touch the film with a ten-foot pole because it told a terribly sad story. He advised the filmmakers to rework the narrative so that it felt hopeful, and then to give him a call.

At the time, I was a wide-eyed aspiring media producer, and I remember feeling incredibly disappointed. One of the most influential environmental foundations in the country wanted to make the film less painful to watch, by pretending that the orangutans' struggle was more solvable than it was. They needed it to end on a note that would send a positive affect vibrating through the bodies of viewers as they left the cinema. Of course, they wanted to see a return on their investment, and I suppose that's what they were looking for with their recommendation of a happier ending. But I thought, doesn't that need to be squared with the reality of the stories they're in the business of popularizing? To get their support, the filmmakers would need to wrap their story in a bow, privileging a human viewer's psychic comfort over an undeniably grim non-human reality. I remember the disturbing realization that this incentive, to produce a palatable story rather than to create new norms for processing vulnerability—the animals' and our own—had likely been behind many environmental productions I'd already seen.

What I saw that day was the manifestation of a tired and longstanding debate about environmental storytelling that pits hope against fear. Per Espen Stoknes, psychologist and author of *What We Think About When We Try Not to Think About Global Warming*, says in a TED Talk, "So many of us are now suffering a kind of apocalypse fatigue, getting numb from too much collapse porn." We need to avoid framing climate stories with disaster, he argues, and instead "reframe climate as being about new tech opportunities, about safety and about new jobs. Solar jobs, for instance, are seeing an amazing growth." He goes so far as to say that all approaches to climate storytelling must use "supportive framings that do not backfire by creating negative feelings." Steer clear of the apocalyptic, Stoknes tells us, and spread the word of "green growth," "happiness and the good life," "stewardship and ethics," and "rewilding and restoration." Yes, fantastic, emphasize the positive—but to the exclusion of everything else? As the external obviousness of how much danger we are in continues to mount, this approach feels flagrantly disingenuous—a narrative denial of reality.

The urge to focus on hope and positive messaging while ignoring what is stress-inducing comes from a desire to mitigate a negative response. Cognitive psychology certainly backs up the idea that our brains over-respond to perceived losses over gains, so difficult information should be balanced with inspiring news, while uncertainty is the trickster in all of this. Uncertainty—about how high temperatures will rise, how nations will act, and how resources will be shared—is embedded in the climate crisis. It is difficult to cope with because the brain has a negativity bias, and this bias can be toxic when it fixates so much on the negative that it blocks out the positive. As the psychologist Rick Hanson describes in his book *Resilient*, our brains are like Velcro for bad experiences and like Teflon for good ones. That's why we so easily obsess about what's frightening to us, and often imagine the worst when facing uncertainty. But as our fear takes over, our creativity, imaginative capacities, and compassion shut down, and we respond from an inflexible position.

When fear is activated, our brain's amygdala sets off an alarm to signal that something is wrong, which generates a response of "fight, flight, freeze, or fawn" to deal with the threat. Depending on the circumstances, the initial fear signal can settle into anxiety, a prospective and future-oriented emotion, which is less intense than fear but can be much longer lasting. As this goes on, our limbic system can become stressed, and put us into an emotionally defensive mode. In defence mode, we can't think critically anymore because we lose contact with the prefrontal cortex, the most recently evolved part of our brain, where executive functioning happens.

Executive functioning is a kind of sophisticated cognition that allows us to do things such as organize our thoughts and actions in alignment with internal goals, think clearly about conflicting ideas, contemplate future consequences based on the present, and moderate our social behaviour with morality. That kind of enlightened thinking is necessary in order to make good decisions and cope well with the long-term effects of a crisis, let alone be our best selves in day-to-day life. But when we are too stressed out, our prefrontal cortex gets effectively knocked offline. We lose access to our capacity to think in the ways we want, and we react from an unproductively rigid position.

A sense of realistic danger is what's fuelling the rise of eco-anxiety. It emerges when we feel our vulnerability and connection to what's unravelling around us, and becomes adaptive when we are in touch with our capacity to care. In this sense, eco-anxiety works like an antidote to the culture of uncare. That's why some call it eco-compassion or eco-empathy instead. It is what happens when we bring our thinking and feeling together—a healthy, human way to function, as long as we learn to stave off its ability to hijack our brains entirely, which later chapters will explore.

Thankfully, my dad did not die that night in 1962. It was a miracle for the world that the missiles never launched from Cuba. My brother

Michael got to grow up and have two children of his own. My dad would go on to meet his second wife decades later, who then became my mom (and he married a third wife, my stepmom, after that). When I was the same age my brother Michael was during the "end-of-the-world party," my parents bought a house in the High Park neighbourhood of west-end Toronto. It had a big, sloping backyard. One day my dad was digging back there, working on the garden, "and then clunk, clunk, clunk, the spade was hitting something," he recalls. "As I dug, the rock just got bigger and bigger. So I started clearing around this huge, hard thing." My dad had finally got his bunker. Too bad it can't protect anyone from the warming we face today.

KEY TAKEAWAYS

» The nuclear threat and the climate crisis have comparable "psychic numbing" effects, with some key similarities and differences.

» Entrenched fossil fuel interests have knowingly confused the public for decades about the dangers of the greenhouse effect, sacrificing population-wide well-being for tightly controlled short-term gain.

» Disavowal is a powerful form of soft denial that many people live with, which involves knowing and not knowing at the same time.

» Multiple kinds of defences can form "perverse states of mind" that actively work against our ability to mitigate and adapt to the crisis.

» The long-standing debate that pits hope against fear in environmental communications is tired, old, and ineffective; it is time to value the wide spectrum of emotions that eco-disruption brings up.

3

DEGREES OF ECO-DISTRESS

They feel incredibly isolated and overwhelmed.
They crave solidarity. They wonder if they're
deviant for feeling this way.
—JENNIFER ATKINSON, climate educator and researcher

Chicken Little is the pseudonym of a thirty-year-old woman who lives in Baltimore. She's struggled with anxiety and depression her whole life, but her mental health really came to a head in 2012, when she was twenty, in the form of a major depressive episode. A few things had precipitated it, including a bad breakup and a move, but she tells me her depression was mostly brought on by her conviction that climate apocalypse was nigh. Climate change as a concept was mainstream in 2012, but "the actual idea that we are all going to die because of it was not yet acceptable or spoken," she told me. In August of that year, she wrote about her overwhelming fears to the

Tumblr advice columnist known as Dear Coquette, asking, "Should I just savor the present moment until I can't anymore?"

Dear Coquette's published response was to the point: "The sky isn't falling, Chicken Little. You've just got a mild anxiety disorder."

When Hurricane Sandy hit New York City in the fall of 2012, disfiguring more than 650,000 houses and causing more than $70 billion in damages, Chicken Little felt almost vindicated. It was not quite *schadenfreude*, but there was a smug sense of satisfaction. "Hurricane Sandy showed that all my talking and obsessing about the climate is not incorrect. So much anxiety is worrying about things that are never going to happen. But this was like, look, the world *is* ending. So to therapists and people who would listen, it was kind of an anti-authority thing, but I was like, fuck you, I'm smarter than you. You think I'm mentally ill, but I'm a prophet!" she says. Then she went on antidepressants and figured out how to manage her fears.

Chicken Little's experience with eco-anxiety is a delicate one to parse out, because she was clearly already suffering from mental health problems, including clinical anxiety and depression. Of course, if you have an anxiety disorder, the climate crisis is only going to make you more anxious. It gives you something very real to catastrophize about. And when Chicken Little talks about climate change now, she still jumps to conclusions that aren't grounded in evidence, such as "the world is ending" and "we are all going to die from it," which connects with her diagnosis. The fascinating thing about Chicken Little, though, is how the world has caught up with her in only a few short years. Catastrophizing is becoming the norm.

She wrote Dear Coquette again in 2018, and the columnist's thoughts around the climate crisis had changed significantly since their first exchange. Her response began, "Dear Chicken Little, I think the not-so-distant future on this planet will be unimaginably horrible . . . I think it would be kinda cool to skip to the other side of the impending extinction event just to see how it all turns out."

Consider that in 2020 an alarming report written by economists at JPMorgan Chase, which was leaked to journalists, said in no uncertain terms that mass starvation, conflict, deaths, and species extinctions will soar as the planet warms, endangering the economy and stock markets, and that "something will have to change at some point if the human race is going to survive." Canaries in the coal mine—people who've worried about this for years, like Chicken Little—are now being newly validated and supported.

So I asked Chicken Little, "What does it feel like to see so many other people expressing eco-anxiety today?" She replied, "As the conversation has become mainstream and not taboo, it has actually lessened my own anxiety and depression quite a bit, because I feel I'm not alone . . . I'm not Cassandra, I don't have to carry the weight of the world on my shoulders anymore, because everybody else now is too. You all get it."

Although eco-anxiety and climate anxiety should not be pathologized, they can appear in healthy or unhealthy ways. Severe symptoms include insomnia, depressive episodes, clinically definable anxiety, substance abuse, self-harming, and difficulty maintaining functioning—especially when faced with news about climate change. There's also compulsive behaviour such as *climate anorexia*, where someone reduces their carbon footprint so much that they act as though it were best if they ceased to exist, and *climate orthorexia*, an obsession with eating "clean" for the sake of the environment. Milder symptoms of eco-anxiety include occasional sleepless nights, sadness, restlessness, temporary paralysis, mood changes, or a striking need to do a single action, such as recycle efficiently.

The Maldives, one of the lowest-lying nations in the world, is already exposed to intensifying storms and, according to the World Bank, might be entirely submerged by rising sea levels by the end of the century. The Maldivian children whom climate-aware psychotherapist Caroline Hickman works with know this acutely, and are feeling its troublesome effects. Hickman created the conceptual frame below to illustrate how eco-anxiety manifests, based on her clinical observations of people from the Maldives, USA, Brazil, Nigeria, Bangladesh, and her homeland, the United Kingdom. The model isn't intended to "fit" people into neat clinical boxes as much as it is meant to support people as they seek to make sense of their own experience and feelings.

ECO-ANXIETY: RANGE OF FEELINGS

Mild	Some feelings of upset, but these are not constant and can be distracted from. People tend to believe that "others" have the answers to the crisis (usually technologically driven solutions), which provides reassurance and relief from anxiety. The feelings of upset can be transient and assuaged. Anxiety can be reduced by a focus on individual and local actions, such as food choices (eating less meat, for example) and recycling. Some defences of disavowal are present (such as minimizing the scientific evidence). There is a tendency to avoid more painful feelings, which may also be perceived as "negative" or "doom and gloom," such as depression and despair. Little disruption in cognition/thinking.
Medium	Feeling upset more frequently (such as weekly) and beginning to have some doubts about the solutions of "others" to the emergency, but still retains a fundamental belief that solutions will be found and that "experts" will deal with things for us.

Medium (continued)	Psychological defences are less efficient in minimizing distress, so there is more strength of feeling and discomfort. Some disruption in cognition/thinking, but not preoccupied by the crisis. Making some lifestyle changes, such as reduction in flying/meat eating, but still choosing to mostly maintain life as before, with minimum significant changes. Can still be reassured by discussion with others. Some knowledge about facts and figures in relation to the climate crisis, but not obsessed—happy to "leave the science to the scientists."
Significant	Daily upset and feelings of distress, with some awareness that this is increasing in duration, frequency, and strength; minimal defences against guilt, grief, and fear. Fears of social collapse can be seen alongside fears about climate change. Anxiety is much harder to mitigate by reassurance. Increase in signs of cognitive/thinking changes, such as guilt and shame in relation to children and grandchildren. Very little faith that "others" will find or act on solutions. Willing to end relationships with people who are in denial about the climate emergency. Frequent feelings of insecurity, but group actions such as activism and campaigning reduce anxiety to a more manageable level. There is a much more significant impact on lifestyle, with people willing to undertake large changes such as choosing not to have children or making a commitment to stop flying.

Severe	Severe cognitive/thinking changes, such as intrusive thoughts, poor sleep, and preoccupation with the climate emergency, leading to a struggle to enjoy any aspect of life because of fears for the future. No defences against feelings of terror. Strongly held belief (which can appear as a certainty) that the climate crisis will lead to social collapse and social breakdown. Anticipation of extinction of human species leading to terror rather than anxiety. No belief that "others" (such as "experts" or any other authority figures) will take steps to adapt or mitigate against climate change. Sometimes may be unable to manage emotional responses (such as episodes of crying or angry outbursts, often spontaneous). May be unable to maintain employment because of the disrupted emotions and struggle to think and concentrate. At most extreme, may be thoughts of suicide or having to kill their children in order to save them from a violent death (social collapse) or death by starvation. One of the only ways that personal security can be found is through belonging to a group and group activism, which is driven by global empathy and guilt rather than self-concern. Severe disruption to other aspects of life, such as paying bills or rent, "because it doesn't matter if I have a pension/home/marriage/job because the world is ending soon anyway."

People who are experiencing first-hand the effects of the climate crisis can run into a severe degree of ecological grief. Seminal research on the condition, conducted by Ashlee Cunsolo, zeroes in on the psychological repercussions of sea ice loss for the Inuit of Labrador. There, in Canada's circumpolar North, where the rate of warming is twice the global average, people are feeling the heat.

The Inuit who live there talk about existential distress as they witness the ice, a big part of their identity, vanishing before their eyes. Now that the ice is not solid for as many months as it used to be, people have less time each year to carry out traditional roles such as hunting and fishing. This culturally defining activity, which their ancestors engaged in for thousands of years, is increasingly foreshortened with each winter season, which is spiritually diminishing. To fill the time that would once have been spent on the ice, and to smother the sense of loss, Cunsolo says, members of the community sometimes turn to alcohol or drugs. And increasingly, if they do try to cross the ice at times when it used to be safe, they risk falling through and drowning. On top of that, unstable weather is changing the availability of vegetation they depend on, as well as the migration of caribou—a vital source of protein for the community.

The Inuit of the communities Cunsolo works with liken the emotional toll this takes to experiencing colonization all over again, as their traditions and ways of life melt away because of a crisis they had nothing to do with creating. Other Arctic peoples have expressed similar feelings. Research interviews with Sami reindeer herders in Sweden revealed that they felt grief towards their future because climate change is "yet another stressor on an already heavily burdened industry and culture." This distress is place-based and relates to real losses.

People with a bit more distance from the immediate effects of ecological change, and with a deep awareness of the planetary health crisis, may easily see themselves represented in the various stages of Hickman's spectrum. I felt deeply seen when I first read her conceptual framework. Memories flashed through my mind as I read the "significant"-level description, making it clear to me that that label was apt for my experience. I thought about the time Sebastian and I were driving on the highway, somewhere between New York and Philadelphia, and I was midway into a sermon about environmental doom when he interrupted me. "I hate to say this, but I really need you to become more aware of how you're talking about the climate. It's gotten really dark, you've been doing it for a while now, and it is really starting to affect me." I caught my reflection in the rear-view mirror. My eyes were reddening. My mouth then started to quiver. It had finally come to this: a time out for my depressing verbiage. "I want you to see how unbalanced you've become," he said. I had lost my capacity to talk about our shared lifetime ahead without overlaying a filter of apocalyptic possibility. Nothing about the future felt safe, and my fear was spilling over onto the person I was closest to.

Pre-existing anxiety sensitivity may play a role in how eco-anxiety shows up in one's life, but it is not a necessary prerequisite. I, for instance, have never had an anxiety disorder. In one study of young people's psychological responses to climate change and COVID, conducted in the UK, researchers found that already anxious youth tended to get stressed by both climate and COVID, but for youth with no pre-existing anxiety disorder, it was more common to get wound up by the climate than COVID. In another study, titled "My Worries Are Rational, Climate Change Is Not," social psychologists found that worrying habitually about the environment did not correlate with pathological worry. In other words, the kind of worrying that is related to diagnosable psychological problems was not an underlying cause of people's ecological concern. "Instead, habitual ecological worrying was associated with pro-environmental attitudes and behaviors, and with a personality structure characterized by imagination and an appreciation for new ideas," the authors write.

So we've determined that eco-distress isn't pathological, nor is it inextricably tied to other forms of anxiety. It's rooted in external circumstances and—as described in an article called "Climate Dialectics in Psychotherapy: Holding Open the Space between Abyss and Advance"—the inability to solve the problem at an individual level or to contain the massiveness of the crisis in one's mind (recall the problem of the hyperobject, discussed in the previous chapter). To deal with this overwhelm, the climate-anxious person unconsciously splits the object of climate change into two opposing entities, and places them on either side of a fence. This translates into black-and-white thinking about "it's all going to hell soon" or "[choose your favourite tool: technology, action, etc.] will save us." They then dig their heels in hard on one side or the other, which helps alleviate *cognitive dissonance*—the uncomfortable feeling of entertaining two conflicting thoughts at the same time. Other unconscious defences, such as denial and disavowal, then help maintain this split. As the authors write, "This split is most famously enacted in the gulf between 'doomsayers' and climate 'dismissives.'" In other words, those who seem cool as a cucumber about the climate crisis may actually be incredibly anxious but are unable to face their feelings.

Clearly, I've landed on the doomsayer side, and I've had to learn how to work my way out of it towards a less dualistic place. Everyone has their own unique anxiety sensor. For me, when I'm digging my heels in hard, my anxiety manifests as a tightness in the chest. For others it might be sleeplessness, change of appetite, visions. I interviewed a young mother named Cecilie Glerup for a CBC radio documentary I produced about eco-reproductive concerns. She described the profound climate anxiety she experienced during her pregnancy as being highly visual. "I had all these catastrophic images in my head, of me running with my child and having nowhere to sleep, or us starving, or him experiencing his parents dying."

"So you were pregnant, and these were the images coming to your mind?" I asked.

"Yes. All the time. It concerned me quite a lot . . . and I didn't really feel that I had that many people to talk to about it," she said. She felt alone and embarrassed by her eco-distress, which made it much more difficult to manage.

Disaster imagery, like that described by Cecilie, came up several times as a theme in my research. Climate justice essayist Mary Annaïse Heglar used to see the world through what she calls *climate vision.* "So I'm walking out on the street minding my own business, and I see a tornado coming down the street. I see flood waters coming up," she explained to me one morning in a Manhattan coffee shop.

"You mean you would visualize it?" I asked.

"No, I could see it. It wasn't like I was concentrating to try to see it, it would just appear. I would see that truck right there turning over, and people running. It's a calm street, you don't actually see it, but I would see it. I would look out of my window and see flood waters rushing up, and no joke, I really did think about animals breaking out of the zoo."

"The Bronx Zoo?"

"Yeah. What would happen if there were a strong storm and giraffes were running around? Or a cheetah? I have a deep fear of snakes, so I was constantly thinking about that too. Like, what do I do? What can I float on? Is it even worth it to float out here? Is it worth it to fight to survive in that?"

"So what did you do?"

"I started running. Equally for my health, but, like, if shit goes down, I'm going to need to be able to run. Honestly, it was more of an outlet for the anxiety. It was something I could do, something tangible. As opposed to just being scared all the time."

Heglar's climate vision began in 2014, when she joined the Natural Resources Defense Council as their director of publications. Being the only person in her organization who had to read every single fact sheet and report put out by each of their teams, she wondered if she knew things her colleagues didn't. "I'd think, if these people know what I know, why aren't they freaking out? Climate change is terrifying! But no one wants to believe it is as bad as it is." Their silence made her start to feel as though she was crazy, and then, over time, she became depressed.

When I asked her what the depression was like, she paused, then spoke slowly. "It does sort of set into your bones. I don't remember not being able to get out of bed. I remember a lot of escapism, though. Running was an escape. I did a half marathon. Never thought I'd be doing that shit. It was just sort of always in the back of my head. I definitely felt isolated, and that there were only a small group of people I could talk to."

Not long after Sebastian made it clear that my eco-anxiety was affecting his own well-being, I was back in his home country of Denmark, where we have a close group of friends who work on climate and sustainability issues. I stayed up into the wee hours of the night with some of them talking about the magnitude of the climate crisis, and for the first time the conversation became deeply emotional. We discussed how we might not have kids, or how we might have kids but would they be able to have kids, how it might already be too late.

The next morning, I was overwhelmed with a profound sense of hopelessness. My existential dread had become too much to bear—had risen to the surface through this impassioned discussion—and, over a mournful breakfast with my friend Märtha in an otherwise delightful café, I could not stop sobbing. I had cried about the climate alone before, but until this moment I hadn't broken down in company, as though someone we both loved was dying. Being the healing company that she is, Märtha brought me to a petting zoo after we finished breakfast, thinking it might make the despair that was dancing in our minds sit down. There were some hilariously fat rabbits there, which forced us to laugh, and that felt cathartic. But my tears returned once we left their tiny shelter.

I walked around the city that day feeling twice as heavy as normal. I was officially in town to moderate a discussion about the ethics of human gene editing at a documentary festival, and hypocritically told myself the flight was justified because it would also allow me to see my friends and in-laws. I eventually made my way to the festival grounds to distract myself from my emotions by checking out the artworks they'd installed. A video piece captured my attention, and I grabbed the earphones hanging on a peg below a large flat-screen TV and placed them over my ears. As the sound rushed in, I was transported to Cuba 150 years into the future. The country had turned into a post-apocalyptic, global warming–ridden society, and for a reason that I couldn't understand, every character in the film was a sex-obsessed hedonist living in squalor. Sounds funny, but all I felt was disgust. It was bestial and disturbing, as though Cormac McCarthy's *The Road* were a porno. I abandoned it, feeling deeply unsettled, and moved on to the next piece—a virtual reality experience about our solar system.

As I pulled the VR headset on and adjusted it over my ponytail, I could still feel the warmth of the last person retained in its thick elastic band. Inside the virtual world, I became a blip in the vastness of the cosmos. I slid planets along their orbits like pieces on an abacus. I walked into black holes. It was all so beautiful and I was mesmerized, but after its three acts were completed, the story ended back on Earth, where humans went extinct. I pulled the goggles over my hairline and noticed the next person in line staring at me. I quickly yanked them off, blotted my tears, and handed him the headset without looking him in the eye.

For some relief, I dipped into a separate exhibition next door, expecting it to catapult me into a less intense headspace. I pushed aside the thick velvet curtain to see what lay just beyond it, and found a wall filled with political posters about the dissolution of the EU and democracy itself, including several pieces about Brexit. I felt a sharp pain in my chest, and then it became hard to breathe. All the artists confirmed that the world is falling apart.

I left the festival and walked across town to Copenhagen central station, where I boarded a train to meet my father-in-law for dinner, without Sebastian in tow since I'd come to Denmark for work alone. Henrik was waving jubilantly from the platform when the train rolled in at Nyborg. After dinner, in a dark, smoky bar, the kind that Danes call a "bodega," he carefully asked me if Sebastian and I were ever going to have kids. He was nearly finished having his own five children by the time he was my age, so I understood his puzzlement. I explained that we would like to, but the ecological outlook was such that the decision was very hard to make, particularly for me. He told me gently that he understands it will likely get very bad but that he wants grandchildren all the same.

I wept again on the train ride back to the city. And though I wasn't envisioning the train catching fire or getting flooded, there I was, stuck in my own kind of climate vision. I could see my mom's face, saddened because I'd not given her a grandchild, alongside those of the depressed, angry, and stressed-out teens I had already interviewed, morphing into a dismal view of what any kid we might have could feel like at their age. I had become that stranger you never want to be, the one crying out loud on public transport.

As is my inclination, I sought out solace from thinkers I admire, and some days later found myself in a YouTube rabbit hole, watching Joanna Macy videos. Macy is an activist and author whose life has been dedicated to helping people plug into the interconnectedness of all living things using a mixture of modern systems theory and Indigenous and Buddhist philosophies. Her words punched me in the gut when she said, "We're talking about the gifts of uncertainty. The courage to feel; feel in the present moment; as you become present to your world, then you feel what you're carrying. Usually, if you're being rushed and hurried out of your mind, you don't bother with pain . . . you try to pave it over, block it down, shut it down, turn away, turn it off. But try as we might, it comes up again. The grief. The outrage. The raw fear. What in God's name are we doing to our world and to each other? And now you are not going to fall to the ploy of the industrial growth society to pathologize that pain. Hear me? Don't let people, therapists or well-meaning friends, try to explain it away in terms of your personal biography or that time of month. It is a measure of your evolution; it is a measure of your humanity; it is a measure of your nobility that you have a heart-mind big enough to see and empathize with the outrage being inflicted on our world and all our relations."

Those words broke a dam somewhere deep down in me, and when it burst, the experience finally taught me the appropriateness of the darkness I was feeling. The grief of all that's being destroyed came and rocked me for a while, but instead of resisting, I let it have me. It was in that moment that I realized that being struck every now and then by overwhelming eco-distress is not only a badge of compassion for the Earth and all life on it, it is to be expected. Mourning is the cost of attachment, so they say, and things are unravelling. Our grief is going to show.

When I left Copenhagen and returned to my then home of Brooklyn, I met up with my friend Nadja at a Mexican diner. She knew I was working on this book, and advised me to put myself into a state of at least some delusion in order to protect my sanity. "If I accepted the full truth of what is happening, I'd not be able to function. I'd just want to kill myself," she said. Though I was not suicidal, Nadja could sense that I needed to be reminded that a bright outlook must be nourished, and that it is my duty to do whatever I require to keep my hope for the future alive.

At that point, my research and thinking about what I could do to help myself cope had not yet gelled. I can remember looking to climate scientists and activists for guidance, and I learned that when you're new to feeling deeply disturbed by this crisis, it can be unhelpful to head straight into those circles. People there are often so resolved about the cause that they say there is simply no time for despair. A lot of prominent figures in the climate space—people who need to set an example—echo that idea with their large platforms, and this can reinforce a sense of alienation and shame around eco-anxiety. The thing is, those activists and scientists have often processed their own eco-emotions; they're just not being reflexive about it, or sharing their internal history of mess and hardship. They were on the other side of a mountain while I was standing at the foot, and I needed some help for my climb.

The following year, I watched the world react to the coronavirus. Mainstream frenzy around the public health crisis left climate in the dark instead of highlighting how zoonotic pandemics and climate change are signatures of the exact same problem, one that must be addressed at the exact same time. This underscored my feeling of distrusting others, which is so common with eco-anxiety. It was deeply frustrating to watch the world acknowledge the Grim Reaper when he arrived in viral form but completely ignore his presence in the changing physics all around us. The message the coronavirus pandemic seemed to be sending was: We'll only deal with the climate crisis when it starts killing the people we love, but even then, we still won't all agree about the legitimacy of the threat, and we'll be ready to sacrifice a lot—old people, poor people, racial minorities. Our leaders will say we're all in this together when they know that's really a lie.

And so, even after I had developed some skills for coping with eco-distress, I still found myself intermittently touching down on the harsher parts of the eco-anxiety spectrum. This is what climate-aware therapist Leslie Davenport described to me as "toggling." It is not that we learn how to cope and then simply stay in that brighter place; rather, we move back and forth between distress over difficult information and states of resilience. Over time, however, we become familiar with the way in which we respond to that lateral motion, which is strengthening in itself.

ONE MONTH BEFORE the pandemic set in, I was in the Vancouver Convention Centre on a clear February day, enjoying the dramatic mountain views from the waterfront. I was there to give a talk on ecological grief at a sustainable business conference. The invitation signalled that the discourse was changing around the role of emotions in environmental work; eco-grief sessions have not historically been hallmarks of business convenings. During my talk, I spoke about the particular kind of grief that arises from ecological loss. I explained that when we lose someone we love, we have socially acceptable ways of grieving. Customs such as funerals, wakes, and shiva allow us to access the social supports we need in what are profoundly disorienting times. But when we're losing species, glaciers, landscapes, stable food and water sources, or zones of habitability, there are no norms or socially organized ways to process these losses.

Dealing with ecological grief is made so much harder because those suffering from it often feel they can't talk about it. If they're kids, the adults at school and at home generally aren't bringing it up. If they're adults, these feelings are still too taboo to discuss in front of colleagues and friends. This gaping silence is similar to what often happens around suicide, abortion, pregnancy loss, and infertility, creating what psychologists call "disenfranchised grief"—grief that is not recognized. Often, it is met with an invalidating response like, "Don't be so dire, you're fine."

While I was up there speaking, a woman in the audience logged into the conference app to send me a direct message. "As someone who frequently cries for the animals, thank you for articulating my feelings to the public. I often feel like a weirdo for being so emotional about it." I saw the message once I got off the stage and responded, "You're definitely not alone. There are some resources out there now if you want to find places to talk about this stuff. I would be happy to share." I didn't know who I was messaging with and tried to guess who this "Alexandra" was, among the women I could spot from my seat. There were so many people looking down at their phones, it really wasn't obvious.

Then a notification lit up. "Do you have time to grab a quick coffee? I feel like I am alone in feeling this guilt and fear, and my sorrow for the animals is alienating me from others." I told her where to find me, and headed there after the session.

When she arrived, we shook hands, but there was no small talk. She was about my age and explained how she had recently stayed up all night reading about the billions of animals that had been burned alive in Australia, and then she burst into tears. It was an amazing moment of vulnerability with a complete stranger, who really needed to know that she wasn't the only one feeling this way.

Alexandra is an environmental professional, just like Mary Annaïse Heglar. Their particularly intense experiences with eco-distress foreshadow what all of us may become susceptible to as the consequences of the planetary health crisis become harder to deny. Similarly, I at various times have felt the urge to quit this research because it is so painful. It alienates me from many people in my life whom I'm very close to but who don't seem nearly as bothered, which is its own kind of torment. I've dreamt of spending my days thinking about something that's far less harrowing and easier to leave at the office. As Panu Pihkala writes in *The Cost of Bearing Witness to the Environmental Crisis*, the most common reactions to the vicarious trauma felt by environmental professionals and activists "include psychic numbing, compassion fatigue, and burnouts."

That's what Jennifer Atkinson, a professor at the University of Washington, noticed among the science students at her school in 2017, when many of them started saying they wanted to quit their environmental programs. It wasn't uncommon to hear these students say, "Why should I go to school when there won't even be a future with a job for me to work in?" To engage with their despair, Atkinson created a seminar for science majors that addresses the emotions of environmental work through the humanities. Disciplines such as cultural studies, literature, history, philosophy, and ethics help us understand our belief systems, our deepest values, and the political structures that created the infrastructure for the carbon economy. They ask questions about how we got into this mess and what social conditions allow environmental injustice to perpetuate, and offer ways to think critically about the world we've built. Crucially, narrative and storytelling are also powerful tools for shaping what we imagine, and in this sense generate new understandings of what might be possible in any future we bring forth.

The first time Atkinson offered the seminar, it filled up overnight, and has run at full capacity ever since. In a free-write activity from the first week of class, the students held nothing back about how they were feeling, writing things like: "I feel hopeless, useless, futureless, angry, and powerless"; "When I think about the future, I feel like all I can imagine is apocalypse"; "I imagine a hellacious nightmare—water wars, mass extinction, infertile soil . . . I have nightmares weekly about our environment"; and "I feel . . . angry, afraid, frustrated, but also hopeless and apathetic."

I asked Atkinson what she thinks the students get out of her humanities course, and she explained how their discussions and assignments work to pierce the bubble of silence around difficult eco-emotions. That simple act is deeply empowering to students, and helps to relieve them of the sense of isolation, alienation, and confusion generated by their intense feelings. The class itself creates a platform for solidarity, encourages flexible thinking, and unleashes new ideas for how young people can conceive of themselves in the world and what they have to offer—all of this within a community of people who "get it."

Fox News ran a segment on Atkinson's class after it launched, which they framed as using tax dollars for "climate therapy for snowflakes." Internet trolls filled her voice mail for months with verbal abuse, and her university had to remove her contact details from its website to block the barrage of messages that called her students "wimpy" and "coddled" babies who needed to "grow up." One person asked, "Do the students roll out nap mats and curl up in the fetal position with their blankies and pacifiers while listening to her lectures?" Their words all revealed a deep-seated belief that reason and emotion are mutually exclusive, which was exactly what the course was designed to transform students' thinking around.

Atkinson still remembers the vivid horror she felt when she first read Rachel Carson's famous book *Silent Spring* in high school. It taught her about the human birth defects and near extinction of the bald eagle that DDT was causing in North America. She still has her first copy, in whose pages her younger self circled language like "strange blight," "evil spell," and "shadow of death," all descriptions of what had hit American towns. Carson was criticized by the chemical industry for conjuring strong emotions in her writing with those kinds of affective terms. They tried to make her out as just another hyperemotional woman rather than a high-quality researcher. But in the end, Carson's meticulously gathered data spoke for itself, she stood by her points, and *Silent Spring* is now widely credited with launching the environmental movement. This taught Atkinson the power of engaging with emotions at the same time as science, dissolving the hermetic seal between them. She could sense that works like *Silent Spring*—stories that contain strong feelings as well as rational thought—would be therapeutic for her students. She was right. It worked.

Her students talked of being steadily transformed by the course and wrote the following evaluations: "My perspective has shifted from acknowledging the horrors ahead of us, to contemplating what internal mechanisms keep me motivated, and deciding what kind of person I want to be in this world"; "The feelings of powerlessness and hopelessness I described are not magically gone, but I understand that it is possible to feel those things and not completely give up hope"; and "This class has taught me that grief must be felt but love maybe more so, and that even in times of absolute disaster the world keeps going and humans help each other."

The eco-distress that so many of us feel becomes a problem when it causes us to burn out, shut down, turn away, lose function, or get stuck in one dire place. But it can also carry enormously practical benefits when we have a chance to process it; it can urge us to gather new information, reassess our life choices, find a deeper purpose, and make important changes that can help bring about justice-oriented societal shifts. As Atkinson's course demonstrates, there are things we can do to take care of ourselves and each other when we're in this state, to make sure we end up on a more soulfully nourishing path—one that leads to deeper connection with others and a vision of the role we have to play in bringing forth a brighter future, instead of the spiral of despondency we otherwise feel.

Dread is a resource floating freely in the air, and it's this generation's job to capture it. In order to do that, the first thing we must do is find a container for our overwhelming emotions. The idea with containment is that any experience, even difficult ones, can be food for thought, and therefore growth and development, if it is processed in a way that helps a person understand their own feelings. The container is the safe place where we do the processing, whether that's a supportive classroom, an activist group, or a conversation with a friend or therapist who gets it. On the critical importance of containment, Paul Hoggett writes, "Are we able to use our feelings or do they use us? Are we able to regulate and cope with them and use them creatively or are we going to be at their mercy?" When we're at the mercy of our feelings, we often aren't aware of it, and we can fall into a variety of worsening states. But even the smallest degree of eco-distress, when harnessed, also has the potential to spur change for the better.

KEY TAKEAWAYS

- » Eco-anxiety isn't caused by pre-existing anxiety, though there may be a correlation with it.
- » Eco-anxiety is not a monolithic thing but rather occurs along a spectrum, from mild to severe, with different symptoms at each level.
- » Anyone distressed by what is happening will have their own anxiety sensor; some, for example, have "climate visions."
- » Coping with eco-anxiety is an ongoing process; we toggle between distress over difficult information and states of resilience.
- » Ecological grief is still, so far, a form of disenfranchised grief.
- » Approaches for containing eco-distress that combine heavy feelings with rational thought (via philosophy, ethics, literature, history, etc.) can be therapeutic.

4

BABY DOOMERS

The egoism of childbearing is like the egoism of colonizing a country—both carry the wish of imprinting yourself on the world, and making it over with your values, and in your image.

—SHEILA HETI, *Motherhood*

Back in 1986, when they got pregnant with me, my parents were having very different conversations from the ones I'm having with my partner now. I wore lace to their wedding that year, draped in my mother's elegant maternity gown. My father wore an ivory pinstripe suit and fedora, markings of the business boom time that it was. By the end of the decade, privatization, free markets, and consumerism would spread like wildfire. Unregulated capitalism blanketed the world.

In 1988, my hometown of Toronto hosted the world's first intergovernmental meeting to discuss climate change and the need for governments, the UN, industry, NGOs, educational institutions, and individuals "to take specific actions to reduce the impending crisis caused by the pollution of the atmosphere." I was not yet two years old. Had leaders from high-emitting countries taken the message seriously, and started lowering emissions then, we could have slowly drawn the pollution down by a couple of percentage points a year. Naomi Klein describes this step that no one took as "a very moderate, gradual, centrist type of phaseout."

Instead, our emissions since that meeting have been climbing ever upwards. The UN tells us we need to halve our emissions by 2030 if we are to have a 50 percent chance at avoiding irreversible climate catastrophe, which means limiting temperature rise by the year 2100 to just 1.5°C. But the world has already warmed by 1.2°C since pre-industrial times, and our current trajectory shows that we could get to 1.5°C as soon as 2030. By then, my reproductive window will almost certainly be closed, and the ticking in my head now drowns out other daily thoughts. Two clocks in sync, one biological, one planetary. Their alarms both set to when we light our world on fire.

As I write this, I'm thirty-three, the same age my mom was when she had me, and there are plenty of reasons why making this decision now is different than it was for her. Along with the climate crisis, there are a number of factors that help explain why many of my peers are hesitating to reproduce: the disappearance of job security, the creeping disintegration of many democracies, and the fact that my generation is the first in modern history not to do better than our parents. As the renowned environmentalist Bill McKibben put it in his climate newsletter for the *New Yorker*: "If you anticipated that your life was going to be punctuated by one major disaster after another, would you be eager to have kids?"

Many of us are balancing the marrow-deep desire for a child against an awareness that as they grow up, there may be no bugs left to swat away when we're walking in the wilderness, no coral reefs to inspire awe, no fresh water available in the United Kingdom, no crops with enough nutrient quantity to ensure billions of people in Africa and Asia are adequately nourished, no ice for Inuit to hunt on, no glaciers to explore, no Tuvalu or Manhattan above sea level. We worry that there will be constant food and water shortages, social division, and wars, and that mass deaths, through both conflict and disaster, will become a casual occurrence from which no geo-location is guaranteed to keep one safe.

Sebastian thinks that most people would rather be born than not, so even if things get fairly Mad Max, we'll never face the day when our child looks us in the eye and says: "You knew all this was coming, and decided to bring me into this situation, when I didn't need to be here?" Personally, I'm not so sure. One sixteen-year-old I spoke with explained that it isn't that she wishes to be dead, but she finds her parents' lack of thoughtfulness astonishing. "What were they doing? Were they not clued in at all? Because I'm so stressed out about my future," she said. Similarly, Caroline Hickman wrote about a small number of her clients who feel rage towards adults and can hardly tolerate being with their parents. One young woman was furious about her parents' decision to get pregnant with her while knowing that the environmental crisis was getting worse. She experienced it as a sign of their lack of compassion, for if they'd cared, they would have spared her the involuntary role of witness to nature's gradual destruction over her entire lifetime. Of all the considerations a prospective mother could now have, this one haunts me the most. However, Hickman says the girl's resentment of her parents stems largely from the fact that they don't legitimize her fears—a sign of their denial and emotional immaturity.

When I described this concern to Sebastian, it unlocked a deeper dilemma—to have or not have kids as a source of deeper meaning and purpose. He said, "Why do people have kids? Is it because of the kid? No. You don't know them. You're having kids for selfish reasons, largely. Imagine if we don't have kids and one of us dies. The other one will be completely, utterly lost. If we have a kid, there will be a sense of a family and meaning from knowing you exist for another person beyond yourself." Maybe I could use this selfish desire as a more enduring motivation to try to make the world a safer place for our child. There's nothing quite like having skin in the game, after all. The faulty element of this thinking, though, is that the climate crisis already threatens to hurt my friends' kids and the young people in my extended family, let alone all the other children in the world. Plenty of my skin is already in.

What I sense more deeply is that I want to have a baby in order to experience the deepest registers of love that humans are capable of feeling. I certainly need one if I want to count tiny toes in bed with Sebastian on lazy weekend mornings—one of the most heartwarming visions in my mind. A big part of me wants to find out what happens in a lifelong relationship with a little person we create. What challenges, highlights, and huge surprises will it bring? Would we feel more continuous joy? Be so moved by our desire to provide a good life for this little person that a new layer of our human potential would be unleashed? Feel deeper registers of satisfaction for what we'd done with our lives? I've never cared much for the idea of legacy, but I can't rule out the possibility that there could also be some unconscious forces motivating me to try to build my legacy biologically. To Sheila Heti's point in ***Motherhood***, there is undeniably an egoism behind child-bearing. The self-serving impulse that undergirds the act of having a child when it is carried out volitionally is perhaps the most common, and thereby least remarkable, form of self-interest our species demonstrates.

Ironically, child-free people have often been stigmatized as "selfish" for deciding to enjoy life without kids. However, this label is increasingly used for people who are deciding to reproduce in the climate crisis. I've found parallels with Caroline Hickman's research among some (not all) Gen Z youth I've spoken with, who expressed a new generational attitude: they believe that bringing new children into this world is unfair to those kids, and therefore speaks to the selfishness of the parents. Another way to look at it is that it takes tremendous courage and hope in humanity's futurity to birth new babies at this time.

At dinner one night, my Gen X brother, who has two daughters of his own, reached out and grabbed my elbow with a pleading grip. "It makes me so upset, what you're thinking when it comes to not having babies. It makes me *so* upset. I just think, how did your climate change research fuck this up so much? You could have a little *you*! Don't you know how valuable that is?" Then his girlfriend cut him off. "Okay, that's enough, leave her alone, she'll figure it out." Clearly, this isn't a private decision; it's one that energizes my whole family.

"Your grandfather used to say, 'It's a race between education and catastrophe,'" my dad told me one morning in a west-end Toronto diner, after he'd just described the shit-kicking he sees humanity taking in the decades to come. "But even so, we cannot give up hope and we cannot give up procreation. That would mean giving up on life, and that would be absolute defeat. We cannot afford to give up our hope in life."

"So you think I should obviously have a child, then?"

"Well, yes. Children give you energy, and I believe that you can do so much more than you think you can, in every respect. My kids are a huge bulwark for me. A huge intellectual and emotional resource. Have faith in the capacity of humanity to survive. And believe in the intellect of your child to survive."

But herein lies the issue as I see it: I want my child to do more than just survive. Can she or he thrive in the calamitous future coming at us?

My stepmom didn't have biological kids of her own, but she helped raise me since I was very young and would certainly be a grandmother to any child I might have. She tells me she feels conflicted. "Part of me understands why you're so worried. It's folly of us to think that things can go on like this forever. But I say you've got to go for it, myself. As long as you're acting responsibly, I think to deny yourself that would be foolish. If it all goes to hell in a handbasket, well, what will be will be."

My mom says she gets my ambivalence but deeply hopes I will have the experience of bearing a child all the same, since nothing has been more meaningful in her own life. This fundamental argument for having a child is the one I most buy into, because my mom has always modelled it for me. But with the stakes as they are, is the risk really worth the reward?

Seeking out and trusting the advice of my elders has always been important to me. I know they know things that I can't yet understand, but on this issue, my parents can't fully grapple with my generation's concerns. To process my feelings, I needed to consult with people my own age. So one day I posted a question on Facebook: "Does anyone feel concerned about having children today because of climate change?"

When I posted that question in 2017, the media were not paying much attention to reproductive anxiety and its connection to the climate. At that point, polls hadn't yet generated alarming statistics, such as the 2020 survey of two thousand Americans which found that 78 percent of Gen Zers do not plan on having kids because of climate change, and 70 percent of millennials said the same. This was before the 2019 survey of 6,500 Australian women that found a third of those below the age of thirty are worried about having a child because of the climate crisis. And it was before the poll of 1,800 American adults conducted for the *New York Times* in 2018, which found that of the quarter who expected to have fewer children than they considered ideal, a third reported this was motivated by climate concerns.

Matthew Schneider-Mayerson, a co-author of a 2020 study of the reproductive concerns of climate-conscious Americans, tells me that their participants ran the spectrum, from people who have five children to people who have had vasectomies and tubal ligations out of love for the planet (and the belief that humans pollute it). Their detailed survey asked participants about their relationships, emotions, families, visions of the world in 2050, and reproductive intentions. Out of 607 respondents, 60 percent reported being "very concerned" or "extremely concerned" about the carbon footprint of procreation, while 97 percent said they were "very concerned" or "extremely concerned" about the well-being of any child of theirs, hypothetical or existing.

Similarly, when I made that Facebook post back in 2017, US Congresswoman Alexandria Ocasio-Cortez had not yet incensed conservatives when she said to her millions of Instagram followers in 2019 "there's scientific consensus that the lives of children are going to be very difficult, and it does lead young people to have a legitimate question: is it okay to still have children?" Pop star Miley Cyrus had not yet made headlines for saying, "We're getting handed a piece-of-shit planet, and I refuse to hand that down to my child . . . Until I feel like my kid would live on an earth with fish in the water, I'm not bringing in another person to deal with that." And Prince Harry had not yet remarked that he and Meghan Markle will only have two children, maximum, for the sake of the planet. I'm not saying that I was prescient; I'm saying that I did not yet have the backing of mainstream culture when I posted that question, and had no idea what kind of responses I might get. Raising the issue felt risky and deviant, and I was bracing myself for a backlash.

But within an hour of my asking the question, dozens of people were pouring ideas into the thread, largely in an affirmative tone, and the conversation continued for days. Morgan Catalina, a curator who programs a conversation series around sex and intimacy called

Hot Takes, saw the post, was intrigued, and reached out to see if I'd be interested in pitching a workshop around this question. That first workshop in 2018 sold out. We've since done several of them, sometimes together, sometimes not, and more than five hundred people have participated.

Given the flurry of chatter around my post, I made a survey asking more in-depth questions and sent it to my respondents to fill out. One person wrote, "As I understand it, we are headed towards ecological collapse on an unprecedented scale, and there will be significant geo-political consequences as it unfolds. It feels unethical to bring a child into this situation." Another wrote, "I don't really have faith that humans can figure out how to coexist in a safe and healthy way (for the planet and our species) before we feel the full effects of climate change." Many of the comments had this same defeatist vibe. But there was also a comment sprinkled with a faint kind of eco-fascism that I'd come to hear a lot as my research evolved: "The individuals not having children due to these reasons are potentially damning the future by leaving the planet to the less socially conscious." Although eco-fascist thinking focuses on keeping migrants out by way of environmental justification, part of it is preoccupied with the "right" kind of inhabitants—the "proper" citizens that should be born, influence the future, and use what resources are left.

Morgan and I once hosted a workshop at SXSW, the goliath of American media festivals, in Austin, Texas. The plan was to give a presentation examining the issue of reproductive anxiety through the lens of the climate crisis then break into smaller groups, where complete strangers would, hopefully, engage in authentic conversations about what they thought and felt. As I scanned the group of twenty-five attendees, a grey-haired man with a French accent leaned over to me and said, "I have three daughters, all in their twenties, and each one of them has told me that I will not be a grandfather because of climate change. It has been really hard for me to understand why." At the end of the workshop, a few participants came up to thank me, including the woman who at one point stood up and announced to the room that she thinks humans are selfish and should go extinct. Then the Frenchman said, "That was very helpful, I can now see this much better from my daughters' point of view." It gave me hope that merciful conversations between parents and their reproductively abstaining adult children are possible.

Of course, some older people already understand our concerns and feel them personally. David Suzuki, Canada's legendary environmentalist and broadcaster, told me that he feels deep distress for his grandchildren. When his daughter Severn was twelve, she spoke with stark moral astuteness about the need to avert ecological breakdown at the 1992 Earth Summit in Rio. "Daddy, Daddy, the chaos is coming," he remembers her saying to him in private. So, years later, when she called to share the news of her pregnancy, Suzuki recalls his response being: "Jesus, Sev, you of all people see where we are headed. It is going to be a tough world to bring a child into." She agreed, and explained that this was part of the point: having a baby meant that every ounce of her fibre would be committed to protecting whatever she could of the environment for her child. This reason for having a baby, which Suzuki had never considered, blew him away. So by the time Severn's younger sister Sarika announced another grandchild on the way, his response had shifted keys.

"They're going to live with their decisions. And my own sense of activity has been motivated by the knowledge that my generation and the boomers that followed partied as though there was no tomorrow. So my life has been dedicated to trying to help the biosphere recover from that thoughtlessness. Having grandchildren now only motivates me more." But, because he is in his eighties, there's an inevitable sadness about all this. "What causes me pain is not only the knowledge that my grandchildren are coming into a very uncertain world, but that I won't be there to help them when I am dead."

At a workshop we once did in Copenhagen, participants took the material and then used it to argue for the need to have babies in the planetary health crisis more than ever. They floated their thesis with a few key points: kids will provide the human ingenuity that's needed to fix this crisis in the next generation; technologies will appear in the coming years that draw carbon out of the atmosphere in effective ways, so those kids will be doubly safe; any child born now will consume less over their lifetime than present generations because they'll be more aware of the issues; and people need to have kids to motivate them to save the future habitability of the planet. It wasn't surprising that there was more optimism in Denmark, where their welfare state supports growing families, than in America. But I was alarmed when one white Dane cleared his throat and said, "Yes, millions will be killed, and that is terrible, but it is nature regulating itself. And it will be better for the ones who are left."

The view that human-caused ecological collapse is merely nature regulating itself can veer into "lifeboat ethics"—the idea that some people, particularly poor people of colour, are disposable, and that this is expected and okay. If hundreds of people capsize after a shipwreck and there's only one lifeboat, who gets to come on board? If everyone clambers on, the whole thing will go down and everyone will die. But if a few people seize a spot before others pile on, they'll brush off the hands that jut up out of the water searching for a grip, and manage to sail away. In the climate crisis, border closures are the way governments flick away arms. This lack of compassion will prioritize the survival of the privileged while endorsing atrocity for the less fortunate, via harmful neglect.

White, middle-class North Americans—perpetrators of the most carbon-intensive individual lifestyles, broadly speaking—face far fewer risks bringing children into the world than residents of the low- and middle-income countries do. That socio-economic reality doesn't make their worries less legitimate; but, as I've mentioned, it does signal the need for a reckoning that many privileged people from high-income nations have been able to put off until now.

In my experience, reproductive anxiety, and the vulnerability it reveals, is a portal towards a deeper search for justice. Not just climate justice but racial justice, economic justice, generational justice—all kinds. They're deeply connected, and can only be addressed if we work on them all simultaneously. And so this pain offers an opening to the only thing that will help us get through this mess: the solidarity that's required to build a shared sense of "we," the most pressing thing humanity needs in order to save itself.

Rachel Ricketts, a Black Canadian anti-racist activist and author who is also a friend, is similarly questioning whether she and her husband will have a child. She tells me "the ticking time bomb" of irreversible climate catastrophe is incredibly tough to think about, because history shows that as the world becomes more perilous, those with the most power and privilege will continue to exploit, extract, and ensure their safety and survival to the detriment of everybody else's. They've been doing that to people of colour for generations, without the context of a massive climate emergency, so what is it going to look like for any child she might have now? "It feels more terrifying than ever," she told me. While part of Rachel feels compassion for how disquieted I am, another part is saying, "Welcome to having to worry about the livelihood of your children. My mother had to do it, my mother's mother had to do it, my ancestors had their children stolen from them and sold. So, welcome."

Even though many are afraid to have children today because of the threat of the climate emergency, procreation has also always been a way to resist existential pressures. Waubgeshig Rice, an Anishinaabe author from Wasauksing First Nation in Ontario, wrote a thriller called *Moon of the Crusted Snow* about climate breakdown on a reserve. Aileen, the grandmother in the novel, believes that the apocalypse they are experiencing is not the real end of the world—for the end of the world has already happened to them, many times before. It happened first when the white settlers came and robbed the land, then when they took their language, then again when they stole their children and sent them to residential schools. "The history of my people is that of apocalypse," Rice told me in an interview.

Anishinaabe people uphold their relationship to the land, the water, and other species as life's most important obligation. And though climate-caused disruptions to this connection are now wreaking ecological grief in Indigenous communities around the world, Rice hasn't heard of anyone from his nation considering not having kids as a way to deal with the changes. "We're in the best position now as Indigenous people to raise proud and aware Indigenous kids," he said. Cultural healing takes many forms, and is inherent in the Anishinaabemowin language lessons he is overjoyed to give his son. Having children to pass culture on to is key to resisting the colonial project that caused their first apocalypse, and many since.

Similarly, the act of having Black children in America has been essential to the project of Black liberation. Jade Sasser, a professor of Gender and Sexuality Studies at University of California, Riverside, who is Black and has also been studying how existential threats affect procreative decisions, told me, "The thing is, we have to assert that we will be here in the future because there has always been an oppressive force designed to tell us we can't be here right now." Saying *the future has us in it too* is a hopeful political position that gets bolstered by having children. Children encapsulate that sense of possibility and a commitment to making sure there are better conditions for Black Americans for generations to come.

At the same time, there are very real reproductive threats that specifically prey on Black Americans and make this form of resistance incredibly hard to navigate. A lot of Black women in the US are terrified of getting pregnant or having children. Not only might their kids experience racialized violence, but Black women are three to four times more likely to die from pregnancy-related causes in the American medical system than white women, a rate that increases to four or five times if the woman is over thirty. Research also shows that Black and Hispanic women in the US, who more often work outside and have to endure brutal temperatures without access to air conditioning, experience higher rates of premature deliveries and stillbirths due to overheating.

Given that reproductive racism is long-standing and pervasive, people of colour have created tools and strategies to resist it. In 1994, a group of African-American women formed the Sister Song Reproductive Justice Collective, and defined ***reproductive justice*** as "the human right to maintain personal bodily autonomy, have children, not have children, and parent the children we have in safe and sustainable communities." Loretta Ross is one of the founders of the collective. When she was twenty-three, she was sterilized by the Dalkon Shield, an intrauterine contraceptive device. In the 1970s, as many as 200,000 American women claimed the Dalkon Shield had harmed them, sometimes causing infertility and even death. This sterilization occurred after Ross had already been forced to have a child born of incest.

These types of reproductive traumas, which disproportionately affect low-income women of colour, are what the reproductive justice framework was created to address. Beyond defending reproductive autonomy, it draws attention to the safety and health of communities. As the latter is precisely what the climate crisis threatens, the reproductive justice framework has been adopted by activists who are concerned about creating families in times of rapid warming.

In 2015, climate activists Josephine Ferorelli and Meghan Kallman created Conceivable Future, which their website describes as "a women-led network of Americans bringing awareness to the threat climate change poses to reproductive justice." Conceivable Future's strategy is to throw "house parties"—gatherings where people can safely explore and express their feelings about how the climate crisis is affecting their reproductive lives, and record video testimonies, which can be viewed on their website. Kallman thinks the phrasing *What do you do when the apocalypse is part of your family planning?* is a good summary of what they're asking, and told me, "We don't have a prescription, we don't have an answer, but we perceive a really clear need to support each other in trying to figure that out." And though they emphasize the importance of emotions, openness, and vulnerability, their larger goal is political. By bringing people together to discuss what they feel, they hope participants will tap into opportunities for taking collective action. Their welcoming approach to parents and non-parents alike is aligned with a sustainable community model; at their house parties, they guide people towards finding ways of supporting green energy and protesting fossil fuel subsidies.

Kallman and Ferorelli were initially a bit uneasy about talking to me because they've been misrepresented on several occasions by journalists who portray them as promoting population control—as though they're waging a campaign like the Center of Biological Diversity's "Get Whacked for Wildlife" initiative, which encourages men to get vasectomies to save the planet. Though its mission is much broader and "family friendly," Conceivable Future's story has often been misunderstood as misanthropic. Kallman and Ferorelli are unnerved by the environmental movement's dark history of concerning itself with other people's fertility, which has in many cases provided justification for coercive sterilizations and late-term abortions and stoked fears about immigration.

In the 1960s, Western thinkers began to promote birth control policies in response to what they regarded as uncontrolled population growth in the Global South. American biologist Paul Ehrlich's *The Population Bomb*, published in 1968, opens with a vivid description of a "stinking hot night in Delhi" and the human presence that caused him much anxiety: "People eating, people washing, people sleeping. People visiting, arguing, and screaming. People thrusting their hands through the taxi window, begging. People defecating and urinating. People clinging to buses. People herding animals. People, people, people, people." The book, which became a bestseller, put forth the idea that the world's greatest crisis was that it was dangerously full. The next year, American activist Stephanie Mills delivered a college commencement address called "The Future Is a Cruel Hoax" in which she declared that she was "terribly saddened that the most humane thing for me to do is to have no children at all" because "we are breeding ourselves out of existence." Referring to the overpopulation issue that Ehrlich had made so popular in the public's imagination, she became famous overnight for her remarks.

By the 1970s, powerful globally operating philanthropic organizations were convinced that preventing more poor people from being born would save the world. The Ford and Rockefeller Foundations were sponsoring the cause, and substantial pressures mounted from other international interests for India to curb its population growth. In *Fatal Misconception: The Struggle to Control World Population*, Matt Connelly explains how during the Indian Emergency period of 1975–77, brought on by the Indira Gandhi government, mass sterilization camps took the fertility of millions of Indians. Whole villages were forcibly sterilized. Vasectomies were preferred because they don't require invasive surgery, but women's civil liberties were also dreadfully curtailed.

China's one-child policy, introduced in 1979 and terminated in 2015, brought about similar human rights violations. So did the forced sterilizations of Mexican-American women at Los Angeles County USC Medical Center in the late sixties and early seventies, and several decades of surgeries to remove the reproductive capacity of Indigenous women without their consent in Canadian public hospitals. The spectre of these historical realities, and several others like them, compels Kallman and Ferorelli to spend a lot of time educating other environmentalists about the dangers of reproductive coercion and the environmental movement's dark preoccupation with it. They make it crystal clear that they do not consider the carbon footprint of each child to be an ecological sin.

But if they had to point fingers, their own country—the United States—would be blushing. Though many racist moments in environmentalist history have focused on curtailing the fertility of increasingly young non-white populations in the Global South and east Asia, it is the consumption of individuals in many of the world's wealthy nations, where the birth rate is now declining, that emits the most carbon. It isn't the industrialized child that is so harmful, but the diapers, clothes, and eventual sense of material entitlement they so often develop. It's not so much about birth rates as it is about how the culture consumes.

One study showed that, on average, having one fewer child in an industrialized country saves fifty-nine tons of carbon dioxide per year, which is 24 times more emissions savings than living car-free for one year, and 37 times more effective than avoiding one transatlantic flight per year. Another demonstrated that a Bangladeshi child only adds 56 cubic metric tons of carbon to their parents' carbon legacy; an American child, on average, adds 9,441 to theirs.

These scientific articles, while they gain a lot of attention, disregard many crucial factors. For example, carbon emissions differ not only between nations but between households within nations, which highlights the importance of consumption patterns and the unequal distribution of wealth, two factors that country-based assessments don't capture. If every generation is 100 percent responsible for the carbon emissions of its children, why aren't our own emissions today simply our parents' fault? According to this logic, your parents probably shouldn't have had you, and then the world would be headed for a safer climate.

Beyond these logical discrepancies, a fundamental human factor is missing when we equate having children with cross-continental air travel or driving a car, and reduce their impact on the world to bar graphs of gigatons—namely, our deeper emotional reason for having them. And the research studies I've just cited, while they offer valuable data, also suggest responsibility for the climate crisis lies with individuals. A more thorough analysis of the issue would include evidence of the towering malfeasance of the fossil fuel industry and a glaring lack of moral leadership alongside the data on personal consumption. Despite flawed arguments about overpopulation and the carbon footprints of kids, however, those who advocate for smaller families still raise an important question: Shouldn't we be mindful of how many kids are being put in harm's way in a warming world?

Some ethicists have used these statistics to argue that parents from nations with huge carbon footprints should think the hardest about how many kids they have. Adoption often comes up in the conversation. If you believe, as bioethicist Travis Reider does, that it is more important to make people happy than to make happy people, adoption allows you to care for someone who's already here and ideally provide them with a good life, without adding one more person—and their emissions—to the planet. When having a kid feels too hard to square with your environmental concerns, adoption can allow you to still have a family.

Shefali Chakrabarty is an environmental engineer living in Australia who opted for adoption out of eco-concerns. After a long and gruelling process, she and her husband eventually brought their adopted son down under from South Korea. Although she now has the family she was longing for, when people say, "There's always the option of adoption," she grimaces. "It's not for the faint-hearted. It is very complex and takes a lot of work." While she hasn't compromised her values, she still struggles with the emotional consequences of raising a child in a worsening climate. "By the time he is old enough to process this [being adopted], he might have bigger problems—like if there's no water or space to grow food," she told me.

Conceivable Future advises people not to feel overburdened by guilt and shame about the carbon intensiveness of having a child, no matter where they live. "You know," Kallman told me, "I could reduce my carbon impact until it was nothing. In fact, I could kill myself and nothing would systemically change. That's the absurdity of winnowing away your own carbon footprint." Instead, they want people like them to ask why it costs so much to the environment just to live. It doesn't need to be that way, and the green transition of the economy they're fighting for, which puts an end to fossil fuel subsidies, helps to disempower the idea that children are more damaging than flights in the quest for a better future.

When I first met Kallman and Ferorelli in 2018, they were the figureheads of the "climate babies" debate, and Conceivable Future was the only entity I could find that was explicitly organizing around the issue at that time. Their pioneering work led Blythe Pepino to seek their advice before launching a related group called BirthStrike in the UK in 2019. BirthStrike grew out of Extinction Rebellion, or XR, a movement focused on civil disobedience as a way to demand that governments tell the truth about the climate emergency. "I went to an XR lecture called 'Heading for Extinction: What to Do about It,' and that lecture really catalyzed me," Pepino said. "I was like, fuck, this is really bad." It catapulted her into a month of depression. The thing that got her out of it was a new-found commitment to fighting the climate crisis with everything she had.

Though Pepino told me she really wants to have children with her long-time partner, she doesn't feel comfortable doing it now. "I'm not really sure that it is worthwhile having a family at this point, to put it quite bluntly. And as an activist, I can't go to prison if I'm breastfeeding," she said. Ultimately, her belief is that if eco-distressed people don't have kids, we may have a lot more time and energy to fight the crisis. Embodying the feminist mantra that "the personal is political," BirthStrikers use their reproductive lives as tools for demanding change, as they will only bear children once they feel that governments are adequately addressing the threat. What would this look like? A few things come to mind: an all-out mission to stop burning fossil fuels and keep remaining reserves in the ground; a transition to a green energy economy; protections for the most vulnerable communities, rooted in climate justice; and the kind of international emergency response this crisis demands—like the one stirred up by the coronavirus pandemic.

Pepino's initial goal with BirthStrike was to target media outlets that would not normally cover the environmental crisis in such intimate terms. She believes that mainstream media obsess about three things—capitalism, women, and babies—and so her provocative disruption to the connection between those things via BirthStrike's message was bound to hit the headlines. And it did, over and over again, across CNN, the *Guardian*, Fox News, the BBC, the *New York Times*, and many other platforms. But in 2020, the group announced that they were changing their name. They explained in a public memo that the words *BirthStrike* "did no end of harm in allowing them to be aligned with the 'overpopulation' topic." They had lost control of their own narrative and were being lumped in with the very same narratives about curtailing fertility that Conceivable Future also rejects. On September 1, 2020, all of BirthStrike's social media accounts were deleted, and the group was renamed Grieving Parenthood in the Climate Crisis: Channelling Loss into Climate Justice. Less catchy, for sure, but also far more accurate in describing what they want their activism to do.

Pepino is a millennial, as are many of the activists she's organized with, but a similar platform was created for Gen Z in 2019 by a young Canadian activist named Emma Lim. Lim is the co-founder of Climate Strike Canada, a network of students and young people who strike to demand climate action, which was heavily inspired by Greta Thunberg's school climate strike in Sweden. Lim was the first high-schooler in Ontario to cut class for the climate.

At the First Canada Youth Summit, Lim organized a mass die-in with other young activists, and used it to interrupt Canadian prime minister Justin Trudeau during a town hall. Not long after, she was sitting in the Ontario legislature to observe a vote on whether the province would declare a climate emergency. Lim sat for hours to hear their decision, while the majority of Conservative members of the provincial parliament didn't even bother to show up until the very end, to vote the motion down. They rolled in late with coffees in their hands, laughing and smiling. Lim looked around at the kids surrounding her, who were now crying, and yelled at Doug Ford—the premier of Ontario, who had voted against the motion—to look them in the eyes as he took away their futures. She recounts that he looked right at her, and laughed.

On the Greyhound bus ride home, she called some friends and asked for their help to build a website. A few months later they launched the #NoFutureNoChildren petition as a way to pledge the government to listen to the science and take action, expressing that if they don't, it will be impossible for young people to feel safe enough to have kids of their own. The pledge caught on quickly. Out of the first hundred people she sent it to, only two or three declined to add their signatures, and just one year later, more than ten thousand had signed on.

Unsurprisingly, Lim and Pepino have been publicly attacked for their views. Former BirthStrikers, for instance, have been called every terrible name in the book, from "nutters" and "libtards" to "not even worthy of being raped." Pepino says the backlash also conjures "even darker stuff, like, 'Oh stupid white women not having babies, they're contributing to the Muslim takeover.'" The sexist and racist trolls targeting these activists make the mistake of assuming that only white women take this stand (Lim herself is biracial). Katharine Dow and Heather McMullen are scholars who have studied the wider "birth striking" phenomenon by attending protests and events, analyzing Conceivable Future's testimonials, and doing participant observation. They tell me the group of people asking these questions has some racial diversity and also includes men.

"It might be more of a class thing than a race thing," McMullen told me. People such as Lim and Pepino who organize their activism around reproductive concerns, and people like me who are thinking really deeply about this dilemma, have a high degree of reproductive choice and control. McMullen and Dow argue that this connects with a particularly middle-class sensibility, which includes access to environmental education and the idea that the choices we make as parents reflect on us as individuals.

Class and education are key to this discussion all around the world. Neera Majumdar, a twenty-seven-year-old Indian journalist based in Delhi, who was born into a higher caste family, told me that she doesn't believe it is okay for her to have a child. That is in spite of growing up in a culture where she has been told her entire life that the greatest success she can achieve is to have a son. Her belief is rooted in the crippling anxiety she feels about her diminishing health prospects amidst ecological breakdown, citing climate change, air pollution, toxic metals, and sewage in the Ganges River. Most people she knows just laugh at her concerns, and tell her she will eventually come round to wanting to marry and have a son. "But I know I won't," she said. "I just don't see a future of me living healthily at eighty. Why should I take someone else's life and do that to them too? I'm anxious all the time. I'm anxious about the world. I don't think it is good for anybody to be going through this and thinking about having children."

In addition to her caste privilege, Majumdar is fortunate to have an understanding mother, which affords her freedoms that many

Indian women, subject to patriarchal expectations and limited education, do not have.

Although there are clearly important differences in how the climate threat affects diverse groups, the climate crisis is starting to create some critical overlaps in the stresses that anyone from any background can now feel. Many young people are incredibly pessimistic about their future and what it is turning into. What does this shared understanding provide in terms of new openings for creating the more just future we want? How can reproductive anxiety in the climate crisis become a unifying platform we might harness to advance justice and climate action? These are critical questions—and opportunities—of our time.

Our decisions about whether or not to have children are not solely individual or personal matters. They are facilitated by the kinds of political decisions our representatives make, what kinds of medical care we have access to, how much debt we carry, what our job prospects look like, and whether there is a safe environment for our kids to flourish in. It is not just that we need to know there will be clean water to drink and safe air for future generations to breathe; we also want to feel confident that our kids will not become traumatized by environmental chaos. At this point, it is not possible to source that confidence from our reality, and so the search for it spooks many.

For a few climate-concerned people I spoke with who are planning families today, a key part of their feeling comfortable enough to have a child came down to mustering the courage to accept that their kid may not get to live out a full human lifetime. They acknowledge that the child they're gearing up for, like anyone, might be fatally threatened by climate events, and know that their child's life will still have meaning anyway, no matter its length. None of us are getting out of here alive after all, and there's never been any guarantee for how long we'll get to enjoy our stay.

One study found that the main arguments for having kids among the climate concerned come down to two things: "the parental investment in environmental politics" and "children as future environmentalists." The idea that having kids will make parents more aggressively pro-environmental is the same argument that David Suzuki's daughter put forth when he questioned her decision to get pregnant. Inherent to this argument is the opinion that non-parents are more likely to surrender to ecological demise. As one research participant wrote, "My partner and I discussed whether or not it's worth bringing kids into this situation. However, not having kids also feels like just giving up." This is not an uncommon sentiment, and one I've felt too. But it is also contradicted by the fact that many child-free activists are absolute trailblazers in the environmental movement.

The second argument that already committed and wannabe parents put forth is more straightforward to grasp. They believe they'll raise future climate activists, pro-environmental voters, and professionals who will contribute to a decarbonized society thanks to the values they'll grow up with. In other words, having kids means birthing eco-warriors, who will contribute to the solutions society needs. A hopeful approach, but it also puts a lot of pressure on the kids to fulfill this lofty expectation.

Many purposely child-free and undecided participants in the study cited their preference for spending their time, energy, and resources on mitigating climate change rather than changing diapers, losing countless days to child-friendly activities, and saving up for their kids' education. The resources they talked about were also psychological. One child-free participant said, "Part of why I decided not to have children is that I'm not strong enough to have kids and fight climate change at the same time." Others talked about not wanting to compromise on being able to risk their safety in order to pursue activism, just as Blythe Pepino told me that the possibility of going to prison makes child-rearing unfeasible. On that note, the undecided and child-free also recognized the political power of abstaining from reproduction. This stance is very effective at making the point, whether to one's family or to elected officials, of how existential the stakes have become.

It's clear that having children has always been challenging for reasons unrelated to the climate, affecting some communities far more than others. But there is now a specific kind of fear about our environmental future that complicates matters further, for all parents. As a species, we will always have kids, no matter how bad things get, and that is in many ways a beautiful and necessary thing. But we ought to also consider that when we have children today, we are bringing them into a dangerous situation, where the interconnected planetary systems upon which life depends have been altered at a scale that humanity has never experienced before. Psychiatrist Lise Van Susteren perceives the societal failure to mitigate climate change as a form of child abuse, and says, "Mental health professionals vigorously endorse requirements to report cases of child abuse. It is a legal obligation, but it is also a moral one. Is it any less compelling a moral obligation, in the name of all children now and in the future, to report that we are on track to hand over a planet that may be destroyed for generations to come?"

This failure can be directly linked to the fossil fuel industry. Fossil fuels and corporate greed have no place in anybody's bedroom, but now we are tucked in with them. The fact that so many young people feel it is our personal responsibility to avoid potential suffering at the cost of getting to know our own children fuels our generational rage, which is particularly acute in environmentally aware millennials and zoomers (Gen Z). Sometimes this rage fuels a punk rock approach. "There's part of me who wants to just say fuck it, and try to live as fulfilling a life as possible, and experience all the dimensions of being a thing on this planet, which might mean having a child," investigative climate reporter Geoff Dembicki, who is in his early thirties, told me. He points out that many important things have already been taken from millennials and anyone younger by people with power, from pensions and job security to affordable housing, so why let them take away the joy of having babies too?

THE PHILOSOPHER ANNA TSING says that our task as a civilization now is to learn how to live in ruins. Does that mean we must also learn to have babies in ruins? Nursing our next generation in smoke and ash? What does the human spirit require in order to live in dark times? "The best answer I could give would be: to take responsibility for the survival of something that matters deeply," writes author Dougald Hine. To him, self-preservation doesn't feel like a good foundation for living a meaningful life. I agree. A better foundation has to include taking responsibility for something that has a chance of being around when I'm gone, which might make some difference and be able to leave this world a bit better than how I found it. It could be a policy, a movement, a conscious business, a community centre—or, maybe, a child. For many communities that have long faced existential threat, resilience gets built by cultivating joy in culturally specific ways—through music, food, dancing, art, spirituality, solidarity. There's lots of ways this could go.

Some feminist and queer scholars have suggested that a good way to find meaning is to expand non-heteronormative joys that don't involve making family by biological reproduction. It's a plea to cultivate the arts of living well on a damaged planet in communities of care and concern. Feminist theorist of science and society Donna Haraway has a tag line: "making kin, not babies." It asks us to reduce the number of people on the planet by creating lifelong bonds with non-biological relatives. "Babies should be rare yet precious," she writes, and asks, "What if the new normal were to become a cultural expectation that every new child have several lifetime-committed parents, who were not necessarily each other's lovers and who would birth no more new babies after that, although they might live in multi-child, multi-generational households?"

Haraway's focus on reducing the human population through forging new kinds of familial relationships has earned her no shortage of blowback from critics who feel deeply uncomfortable being told by a white American woman that people anywhere should have fewer children. Some sense a misanthropic sentiment in her words, and critique her argument for "kinnovating" as being based in vague presumptions rather than any evidence that it could really help our ecosystems heal. This debate, a refreshed version of old population wars, will likely see no end. But Haraway's proposal makes the important point that we should spend time reimagining what it means to create nurturing environments for children in the planetary health crisis. At no point in history have people experienced this kind of civilizational change. Why shouldn't strategies for coping with this unprecedented situation include radically different visions for how we create and build family, which the queer community has always led the way on?

Matthew Schneider-Mayerson, quoted earlier for his study on the reproductive concerns of climate-aware adults, explores the particularly high-pressure way of thinking about family in the West in an undergraduate course he's developed called Reproduction in the Age of Climate Change. Through a Western lens, having a child—biologically or via adoption—will demand a bottomless pit of resources and time for at least two decades, but not having a child will condemn you to the life of a lonely spinster. "This is an extremely specific conception of kinship, and it's so binary, impoverished, and isolating," he told me. In his class, students learn about Eastern family structures, Indigenous kinship, queer chosen families, and multispecies families. The readings urge students to think broadly about how to cultivate networks of care and support beyond a nuclear family structure. While he sympathizes with the importance of asking the question "Is it okay to have a child in the climate emergency?" he believes this line of inquiry must be altered. "We should all be engaged in raising children whether or not there's a child with our DNA. To me that is likely to be a more healthy, more just, and also a more joyful world," he said.

One night, while Sebastian and I were playing with our Beautiful Trouble card deck, he asked me to think of any question that was on my mind and then pull a card for an answer. I asked, "If we have a child, is that child going to grow up to feel increasingly stressed about the world and their own life as environmental disruptions get worse, becoming so severe that I'm not wrong to be wondering if it is even okay to have them in the first place?" I pulled a card and flipped it over. It read: THE PROBLEM IS INSIDE OURSELVES.

KEY TAKEAWAYS

» Reproductive anxiety is a rising phenomenon in societies around the world.
» Child-free people have often been stigmatized as "selfish" for deciding to enjoy life without kids, but this label is increasingly used for people who are deciding to reproduce in the climate crisis.
» Overpopulation narratives can be harmful and historically have threatened the civil liberties of Black, Indigenous, and People of Colour.
» There are two central concerns about reproduction among the climate-aware: what having a child will do to further strain the planet, and what a warming world will do to the health and safety of that child.
» There are key differences between how marginalized and privileged groups experience reproductive anxiety due to an increasingly hostile environment, as well as clear overlaps that create opportunities for solidarity.

- Activist organizations have emerged in recent years demanding change by way of abstaining from reproduction until leaders responsibly address the climate crisis, thereby pointing out the existential stakes of inaction.
- Theorists have argued that the climate crisis calls upon us to model alternative family structures and broader support networks.

PART TWO

CONNECT INWARD TO TRANSFORM ONESELF

5

STANDING IN THE SHADE OF THE CAMPHOR TREE

The emotions of climate—the denial, the betrayal, the weight, the gaslighting, this is the story of my life.
—TAMARA LINDEMAN, The Weather Station

In the early days of the COVID-19 pandemic, Charlie Glick, a musician in his late twenties living in California, was strolling through LA's Atwater Village neighbourhood, thinking about work. Music was all he had ever wanted to do with his life, and before the pandemic, playing with his band had been starting to stabilize into something that looked like a career. COVID-19 upended all that, though. Lockdown and social distancing measures meant the band couldn't go on tour or play live shows for who knows how long.

Charlie had always loved camphor trees, and on that day's reflective wander, a remarkably large and friendly-looking one, rooted at a corner on Edenhurst Avenue, beckoned him over to it. He walked under its arms as they rustled in the breeze, and the shade the tree cast over him conjured a sudden intuition that made his blood run cold. "I just had this instantaneous feeling like, oh, the rest of my life is going to be this series of increasingly dire crises," he told me.

It was in that moment, under the camphor's leafy dome, that Charlie understood what many public health officials have said about the pandemic: it is a sign from the Earth that we are rubbing up against ecological limits, and a warning of much worse things to come. Whereas experiencing the climate crisis often meant processing *warnings* about ecological breakdown, living in a pandemic caused by a zoonotic virus *was* the ecological breakdown that climate rhetoric warned about (since then, the hypothesis that the virus leaked from a lab has gained ground, with no conclusive evidence yet for either source). Whether the tree whispered this to him or it all clicked in that moment for a more rational reason doesn't really matter; the result was that the pandemic and the climate crisis ceased to be separate concepts in his mind. One all-enveloping hazard foreshadowed the other and yet was simultaneously indivisible from it. Realizing this sent him spinning down a rabbit hole of grief and anxiety, where he imagined the gritty pain of climate disasters, dwindling energy supplies, political turmoil, and even more pandemics that would punctuate the rest of his life. He felt himself collapse—emotionally and physically—in the shelter of the tree.

"My whole idea of my life was gone. It was really traumatic and everywhere I looked, I would just see fossil fuels. I would see myself, literally, as a product of fossil fuels," he told me. Charlie's hope of being a successful musician relied on tour buses and planes and the countless gas tanks they'd empty, and imagining the pollution from each gig quickly took the shine out of that dream. And the more he thought of himself and the people around him as fleshy fossil fuel products, the more intolerable it felt to live in American society.

Charlie spent the entire summer of 2020 reading, thinking, and talking about ecological and societal collapse to anyone who'd listen. His bandmate Chris told him that his doomsaying was pointless because people won't change until catastrophe hits them personally. This incensed Charlie. "Dude, what do you think is happening? Why do you think I'm going through this? There's a fucking global pandemic!" he replied. Their disagreement was putting serious strain on the band, and over time, Charlie couldn't bring himself to work on music anymore. The idea of being a rock star felt absurdly unimportant in a world on fire. He told his bandmates that he needed to take an indefinite break to radically rethink his life.

Charlie's turning away from the band right as they were finding success was an inexplicable move to everyone who knew him. It was cause for real concern. His personality seemed to have changed overnight, and although his bandmates were very angry with him for pulling the plug on their project, they were equally worried about his mental health.

It wasn't only his bandmates who were upset. Charlie's father threatened to fly out from Delaware to shake some sense into him. In a mocking tone, Charlie recounted to me the phone call where his dad insisted he stay with the band and claimed Charlie's fears were out of proportion with reality. To which Charlie said he was sufficiently educated on the subject of climate change, and then hung up on his father for the first time in his life. As a white American baby boomer, Charlie's dad grew up in the nation and era arguably marked by the greatest abundance of material wealth derived from extracting natural resources that the world has ever seen. Charlie thinks it is nearly impossible for men like his dad to accept that there is no way to sustain the world as they've known it or that massive change is inevitable. So there they were, father and son, unable to speak to each other about their pain—Charlie's on the surface and his dad's just beneath it—making his decision to step away from the band all the more stressful.

Without playing music, Charlie had a lot of time on his hands. He filled it by reading things like the 1972 *Limits to Growth* report from MIT that simulated the dire effect on Earth's non-renewable resources of exponential economic and population growth, which people still debate. He also read *Deep Adaptation*, a 2018 non-peer-reviewed and controversial paper by a professor of sustainability leadership at the University of Cumbria, England, named Jem Bendell that gained a large following for arguing that near-term societal collapse is inevitable. Both publications spelled out the end of the world, and they both felt impossible not to take seriously, despite the misgivings around them. Charlie also scoured the headlines each day for climate news, and read the writings of people expressing their personal climate grief. As his view of the future narrowed with each reading, and his obsession with collapse stories grew, he found himself in a very bad place, and one day he could no longer get out of bed. This went on for some days. That's when he knew he had to do something to help himself.

Charlie had learned about the Climate Psychology Alliance in the UK from his readings, and reached out to them with the hope of finding someone to talk to. They connected him with a climate-aware therapist, and after talking with her for the first time, he felt noticeably better. "We immediately had this connection and it just felt so good to talk to her and feel like I wasn't crazy, because nobody in my life was ready to talk about this stuff," he told me.

"What's the most helpful thing that your therapist has done for you?" I asked.

"The single thing that has really helped me the most is that she told me, 'You have to find other people to talk to, you have to build community.'" His therapist was worried about the way he'd been educating himself about all the worst outcomes. He was cramming in tons of frightening readings within the span of a couple of months, and he was doing it all alone. In contrast, she was in her seventies and told him that it had taken her decades to internalize the same dire material, which allowed for a slower and more balanced type of intake. She urged Charlie to be very careful about his digital diet and find others to speak with who "get it."

He partly followed her advice. He couldn't seem to crawl out of the tunnel of reading terrifying climate news stories and analysis about collapse, but he did take action to connect with local chapters of Extinction Rebellion and the Sunrise Movement, both prominent climate activist groups. Rather quickly, the personal connections he was forging through activism lessened some of the pain, as did his blossoming romance with a woman named Evelyn, who understood and accepted his concerns, even if she didn't feel them as acutely herself.

Pretty soon, things started opening up. He could easily get himself out of bed, was having fewer breakdowns about the climate, and was able to better manage his emotions. For instance, when he and Evelyn took a trip to visit his sister in Chicago who'd just had a baby, he was able to button up the "Doomsday Charlie" side of himself, to not existentially stress his sister out about the fate of her newborn. This was great progress, but he was still struggling despite his growing resilience.

Via the digital channels of Extinction Rebellion LA, he came across an article I published on August 5, 2020, in my newsletter *Gen Dread*. The article was called "Why activism isn't *really* the cure for eco-anxiety and eco-grief." In it, I wrote about what difficult climate feelings call on us to do, referring to the famous saying from folk music legend Joan Baez, "Action is the antidote to despair." In my view, quelling despair isn't quite so simple.

It's true that when we act on our values, we put our core beliefs about how we ought to be in the world into practice, which can bring relief. Climate psychologist Renee Lertzman argues that enormous swaths of the population now feel considerable psychic pain because of the large gap between our environmental values and our actions. In other words, as we continue to work jobs that pollute, buy food that's unsustainably produced, or benefit from policies that privilege the health of some communities while stomping on others, we are living *out of alignment*, and we suffer—often unconsciously—for it. Narrowing that gap through activism is an effective way to feel more at ease.

As eco-anxiety and eco-grief have taken hold of society in new ways over the last few years, the tendency to prescribe action as a tool to beat back the feelings has grown. But climate-aware psychotherapist Caroline Hickman argues there's a danger lurking in that sentiment. It's a shortcut—a too-quick move from pain to action—and it threatens to leave people far less resilient and capable of facing the ecological crisis than they ought to be. It also supports the disenfranchisement of grief, discussed in chapter 3, and mutes expressions of pain in favour of forward momentum.

To fully process these complex feelings, we must move away from the positivist psychological framing that sees some feelings as bad and others as good. Despair and fear are not inherently bad. Hope and optimism are not inherently good. In a course for therapists treating climate-anxious clients, Hickman noted that there are times to be cowardly and to recognize that it takes courage to be so. We must move from an either/or to a both/and model. There is meaning in every emotion.

Hickman says we need to not only *grow up* in the climate crisis by cultivating our imaginative, creative, determined, and hopeful capacities, we also need to *grow down* by building our tolerance for guilt, shame, anxiety, and depression. After all, life in an ecological emergency is not a linear progression. There are uplifting wins and, more often, crushing losses. We need to be able to flexibly bear both by growing up and growing down, so that as we move forward in life, we become deeper human beings.

Notice how, in the diagram below, the person who sees that "This can work and be good" is moving forward with more strength only after having been afraid, guilty, depressed, and so on. The next time they feel a huge hit of loss, they'll descend again. But it will be into a familiar place that lacks the power to swallow them whole, as they'll know from experience that they'll be able to rise again.

The Process of Transition

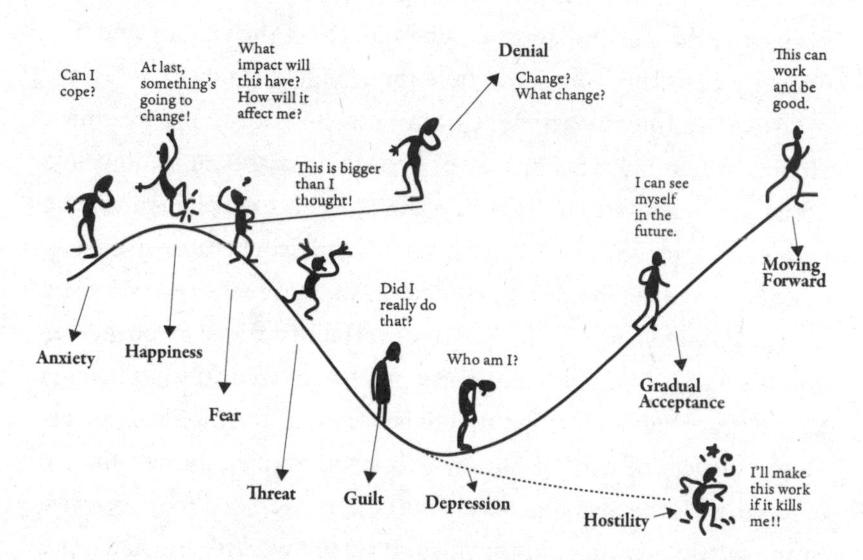

© John Fisher Personal Transition Curve concept and content 2000-13; originally presented at the Tenth International Personal Construct Congress, Berlin, 1999.

And herein lies the key to why it is unhelpful to say that activism is the cure for eco-anxiety and eco-grief. When we're looking for an antidote to pain, we're looking for "happiness" or what we think of as strength. But that impetus tries to cut straight across *the process of transition* from the moment we feel fear encroaching to the place of moving forward. It refuses the painful process of integrating difficult emotions into our life that emotional intelligence requires. It is a flimsy kind of security that bounces back and forth between the onset of fear and the ideal of being a despair-free activist. Eventually, that elastic path will become less flexible and snap or burn out.

We all need to process some of the anxiety, grief, and depression that come with this entirely threatening situation, and learn how to fold them into our lives. This is what Hickman calls *internal activism*, and it is just as important as external activism—the more conventional kind. The trick is not to get lost in the dark places that internal activism brings us to—to keep moving—and to welcome the idea that we'll cycle through the trenches again, because the climate and biodiversity crisis isn't going anywhere for a long, long time.

After reading my article, Charlie reached out to me by email. He described to me his history of eco-distress as well as how he'd jumped into activism to help alleviate it. He also explained that he had recently had to take a step back from activism because of exactly what I'd described in my newsletter. Like many people who've looked to external activism as a "fix" for internal pain, he dove in too quickly and tried to cut straight across the process of transition to being a happy, resolved, and resilient activist. He had confused the therapeutic effects of community, which his therapist wanted him to explore, with the idea that action in the direction of a more positive future would take away his anguish. It wasn't working; he had more internal activism to do.

Don't get me wrong—external action is absolutely vital. Society needs a lot more of it, and contributing to that momentum can bring some genuine calm because it means you're addressing the thing that is stressing you. But "action is the antidote to despair" can oversimplify a complicated experience and indicates a society that is averse to difficult emotions. As eco-anxiety researcher Panu Pihkala writes, "In (over)emphasizing action, one can also see traits that stem from a general avoidance of emotions or even a culture of belittling." In many Western nations, where mental health problems are soaring, we tend to snuff out our feelings by working (too) hard, busying and distracting ourselves, retail therapy, eating and drinking too much, taking drugs, or explaining away our emotions with reason. This is the emotional immaturity of many modern societies at work, which will go to great lengths to block out the deep internal and collective external work that is required to face and process tough feelings to completion. Charlie came to understand this, and rather than trying to paper over his feelings with action, he took more time to sit with them. I was curious to know what he was learning about himself from this process, and so we set up a time to talk some months later about how he was coping.

Activism, he had realized, was just one way of accessing other people who "get it," which his climate-aware therapist rightly urged him to do. When he started reaching out to various writers online who were thinking about these topics (not just me) and having meaningful conversations with them, he quickly cultivated connections that could contain his deepest fears and frustrations. Each authentic conversation about the emotional toll of human-caused destruction of the environment made it all the more bearable, he said. Research backs up what he found, and shows that social support of this kind is vital for sustaining psychological health.

What he was still struggling with when we spoke was a burning need to extricate himself from industrial society as soon as possible. "I'm so unhappy with the way things are, but there is nowhere clear for me to step to," he told me. "We can choose between this horrible, entitled way of life or a more reparative way of life. In theory that sounds great, but there is no road map that I've found for doing that." At a loss for how to practically forge ahead in a way that's more aligned with the needs of the Earth, Charlie found himself at a new crossroads, where two unanswered questions wouldn't let him pass:

"How are you going to live with this knowledge? What are you going to focus on?" The voice inside his head urged him to make meaning from his suffering.

He felt enormously tempted to skip town, move to the woods, and learn to live off the land. Away from the bustle of the big city, he at least wouldn't have to be painfully reminded that the water coming out of his tap was being pumped by fossil fuels from a reservoir three hundred miles away. But then he'd think of all the children in LA who've never had the opportunities he had as a kid to chase their dreams, play music, and just enjoy being young without the spectre of the climate emergency and a pandemic hanging over their heads. Wouldn't he be abandoning an entire generation that needs help fostering their resilience if he just checked out of society and hunkered down for the apocalypse? Responsibility was on his mind. What, after all, do we owe each other at the end of the world as we've known it?

The solution to his predicament wasn't yet possible to know. The answer for now was to take more time—time away from activism, time away from expectation, and time towards experiencing all the individual feelings in the moment as they came and went. He took some months to do this. Then, one day while he was out on a walk, a new realization suddenly clicked (a familiar story by now). "Oh! I should move in with Aunt Wilma so she won't have to live alone!" Charlie had always loved spending time with elderly people, and felt there could be a deep exchange between himself and his ninety-four-year-old aunt in these troubled times. He sat on the feeling for a couple of days before calling her. When he finally did, Aunt Wilma gladly welcomed the idea. So he packed his bags for Delaware and took Evelyn with him.

After he settled into his new life, I emailed with Charlie again, and it sounded as if he was bearing the storm of his despair even better. "Whether I decide to keep doing music or go learn organic agriculture almost doesn't matter to me anymore . . . I just want to do what brings me joy, instead of caring for the Earth out of some lofty feelings of morality or this 'we're going to starve or die in armed conflict if we don't all change our way of life ASAP' mindset that was plaguing me." Speaking of the pandemic and people's urge to "get back to normal" rather than transform our relationship to nature, he added, "It still drives me crazy how the vast majority of people want to get right back onto the self-destruct program, but I'm not sure that's a healthy place to live from, constantly worrying about all of that . . . It's something I can manage and not let rule my life." The time that Charlie had taken to understand his feelings was starting to bear fruit. He was learning to live with the planetary health crisis and accept its gravity with some lightness, in his own way.

Though external activism would have been a noble pursuit into which to pour his energy, alongside internal activism, for now Charlie had found meaning in a much quieter act. Moving in with his aunt while her time on Earth was dwindling was the first step in an endless march towards living more comfortably—more meaningfully—in his awakened state. His tale is far from over; more emergent actions await.

We can learn a lot from Charlie's story. Firstly, it demonstrates an almighty truth that I also realized from my experience with eco-distress: learning to live with it can be a very uncomfortable process and take several months or years. It is an act of labour. It is real emotional work. It can affect your relationships. It can change what you do for work, where you live, and how you spend your days. As Panu Pihkala once told me in an interview, "We need to have enough energy, time, resources, and support to process these existential questions and emotions in order to harness them." We must essentially have our basic hierarchy of needs met, so socio-economics plays a role. In this sense, the ability to live well with eco-anxiety is a justice issue in itself.

Several climate-aware therapists shared with me that their clientele tends to be white, middle-class, university-educated types who are overwhelmed by the frightening nature of what they know about the environment. Therapy is expensive and therefore off limits to many people, especially those most exposed to the hazards of climate and environmental change—namely, people of colour and poor people.

Jennifer Mullan is a clinical psychologist who focuses on decolonizing therapy, which means using alternatives to the mainstream mental health model that can further emotional wellness on a larger collective scale for communities of colour. I wanted to know what she thought the climate-aware therapy field can do to make its services more accessible to the most vulnerable communities, so I gave her a call. Her own clinical practice is centred around the fact that the standard for therapy that's widely in use today, which happens one-on-one and at high cost, was built from a colonial and individualistic biomedical perspective. "The mental health industrial complex, the way that it is set up, continues to serve the elite, or at least the middle-class white person," she said.

Front-line communities may be better served in group therapy in community centres, or in one-on-one therapy at low cost. Honouring ancestors and spirit—as Mullan puts it, "We're going to need to hang on to something outside ourselves," whatever that may be—is a meaningful practice in various cultural contexts. Religion and spirituality may factor strongly into one's world view and coping strategy—anchors that "mainstream" therapy isn't always comfortable addressing. However, climate-conscious therapy is in a good position to dismantle the traditional clinical model—it's already doing so in some ways—and to lead towards a more equitable, multi-dimensional approach to mental health. De-pathologizing eco-distress and treating it as a collective experience are major facets of this shift, alongside an interest in uplifting community support. The Climate Psychology

Alliance, for instance, hosts Climate Cafés—human-centric, emotions-friendly group meetings where people can safely express what they're sensing about what the climate crisis means, not in some far-out future way, but for their own lives and loved ones. They are relational and permission-giving spaces that help people work through their fears and frustrations together.

Along these lines, Charlie's therapist did indeed guide him to start thinking of himself less as an individual who was experiencing a kind of ending, and more as a chink in a long, inter-collective chain of life spanning from his ancestors to future generations. This made him think about his maternal grandparents. They were Mexican and grew up in the desert with absolutely nothing, he told me, and yet they lived meaningful lives. Maybe there was some resilience to be mined there. Maybe he needed to think less about the future and more about the past.

Charlie's first tour down the rabbit hole of eco-anxiety and grief under the camphor tree is an example of how this form of emotional grappling is relational. For Charlie, he was relating to a changing sense of the future itself, and the disappearance of the vision of a relatively disaster-free life that he'd grown up to expect. As with any kind of grief, bidding adieu to the stories you used to live by will carve your stomach out, and that is never not disruptive. Psychologist Ginette Paris writes that the psychic space between old stories and new ones in times of transition "often feels like a deadly zone." It beckons primal emotions and existential feelings about our security, identity, and place in the world.

Eventually, through mourning, our dwelling in this deadly zone and learning to say goodbye makes room for new, nourishing narratives to live by. We need help finding and creating those new positive stories, which is partly why his therapist's advice to build community with people who "get it" was so crucial. Even more so, she knew that Charlie needed to find people who could tell him, "I too have felt myself in that deadly place, and I am still alive." They'd be able to show him how the only way out sometimes is through, and that it is indeed possible to find a way through the most intense forms of this distress.

Coping is a process of learning to live with change, and it will take time regardless of how you may try to rush it, deny it, or push it away. That's because a lot of re-worlding is going on when we practise internal activism as Hickman explains it, or learn to integrate difficult emotions into our lives. Internal activism teaches us how to exist in a place of obliteration that we thought we could not stand and then come to realize we can actually tolerate. It's what the environmental scholar Blanche Verlie calls ***bearing worlds***, which we do by enduring the pain of the end of the world as we have known it and "laboring to generate promising alternatives." Bearing worlds as "internal activists" involves pupating, and when we emerge transformed, we embody new kinds of existential resilience, acceptance, and courage to face our reality.

Now that we are strengthened and more flexible, these attributes hold us back from falling apart in the disorienting ways we used to. That's not to say we won't fall apart again one day in the face of devastating loss; after all, in an ongoing crisis, the task is to master the art of toggling between distressing information or experiences and ways to bear them. But over time, we get familiar with our own responsiveness, which is a kind of emotional intelligence. This translates into the awareness and confidence that we'll be able to move through painful emotions and not get trapped in any one barren place, which creates openings where some hope can be found. For Charlie, hope looked like the realization that his despair was something he could manage and not let rule his life. Who knows where that hope might lead next?

To make sense of Charlie's story and those of so many other eco-distressed people, we must recognize that waking up to the climate and wider ecological crisis, and our radical vulnerability within it, is itself traumatic. Trauma is not merely a symptom of climate change that sets in physically or mentally after an upheaval like a hurricane. Rather, climate change and ecological breakdown—ongoing threats to life that they are—create their own order of trauma and become a new lens through which we see the world: the "climate trauma" lens. This lens can affect our identity in the most fundamental ways, as well as how we make sense of the world. As we've seen, it can determine whether or not we'll have children, how we parent, what job we work at, and what spiritual guidance we seek; and it changes what we notice—like the real story behind a headline about a migrant caravan, or the haunting shade of brownish-green grass on a dry mountainside.

Many people have defined *trauma* over the years. One useful definition comes from Judith Herman's *Trauma and Recovery*, which describes traumatic experiences as those that "overwhelm the ordinary adaptation to life" and "the ordinary systems of care that give people a sense of control, connection, and meaning." Often, the trauma that people experience comes from something bad that happened to them in the past and which they then get stuck reliving, such as a sexual assault. Healing from that kind of trauma is fundamentally different than healing from a trauma like the climate crisis, which is ongoing. For this reason, climate-aware psychotherapist Dan Rubin thinks of climate trauma as more like the cultural trauma of sexism or racism, which actively continues to harm a person over their entire life. Cultural trauma happens to collectives of people, although it can bear down disproportionately on individuals. Researchers Robert Brulle and Kari Norgaard define *cultural trauma* as "a social process that involves the systematic disruption of the cultural basis of a social order." It takes what was once assumed and flips it on its head, affecting our relationships, beliefs, routines, institutions, and general ways of being. Because the climate and wider ecological crisis unceasingly composes and recomposes traumatic experience, the key is not to get over this trauma—for that is not possible as long as temperatures are rising and the mass extinction of species marches on—but to find ways to cope and live better with it.

Once we begin seeing through the climate trauma lens, we are reminded several times a day of all the ways our dominant systems are harming us, and how our decisions that maintain the systems currently in place hurt others as well as ourselves. What is revealed is an economy of pain. As ecopsychologist Zhiwa Woodbury writes, "Climate Trauma is the elephant in the room of every human interaction" that "challenges *the very idea* of a shared future in a way that fundamentally indicts our *present* identity as a species." Once we recognize the power of the climate trauma lens, we can better understand why Charlie felt a desperate need to exit industrial society. That elephant is not native to off-grid communities in the American woods.

Part of what Charlie found attractive about learning to live off the land is that it would relieve his conscience of the guilt, frustration, and disgust he felt for participating in a system that perpetuates climate trauma. He was suffering from moral injury. *Moral injury* was first used to describe the agony soldiers experience when they reflect on memories from combat where they transgressed their own moral boundaries, as well as register that they were helplessly caught up in a vast war machine that hid the truth from them. Climate-aware psychoanalyst Sally Weintrobe says that moral injury is another form of ecological trauma. She argues that we cannot help but feel our souls have been wounded once we grasp how nonchalant our leadership has been about us and our children dying, by continuing to perversely protect the fossil fuel industry despite decades of mounting scientific evidence that pointed to its dangers.

Just as climate anxiety is a sign of compassion, feeling morally injured in the climate crisis is a healthy sign that your conscience is sound and your sense of being transgressed is alert. That's why getting enraged about what is happening can feel good and right. It is the same for any injustice, where in our rage we can bond with others who feel it too, and draw strength from its emotional power. Interestingly, one study that looked at eco-anxiety, eco-depression, and eco-anger found that anger is the most practical and helpful emotion of the three. Some psychologists refute this trend of sticking "eco" on every emotion, when the terms make just as much sense without the prefix. Still, this study found that anger in the context of environmental crisis "predicted better mental health outcomes, as well as greater engagement in pro-climate activism and personal behaviours."

There is a crucial difference, though, between experiencing healthy levels of anger from moral injury and unhealthy degrees of guilt for being caught up in a system that crosses one's moral boundaries. Many citizens of industrialized nations with big carbon footprints feel at least partly responsible for the catastrophic state of things, and of course to some extent we are. But tearing our hair out over the climate cost of each consumer decision is a massive energy drain, and it aims our weapons at the wrong target. Research has shown that just one hundred companies are responsible for 71 percent of the world's greenhouse gas emissions since 1988, and more than half of global emissions can be traced to just twenty-five fossil fuel producers.

We need to learn to stop judging ourselves so harshly in this crisis, from multiple angles. Climate-aware psychotherapist Andrew Bryant told me, "People can get at themselves and say, 'Oh, I shouldn't be feeling sorry for myself, there are other people who are worse off than me' or 'Why don't I just get off my ass and do something?' That inner judge has a big impact on people's sense of capacity." He spends a lot of his time helping people become aware of their inner "environmentalist critic" that tells them that their sense of ineptitude is not okay. In order to move towards meaningful action, we need to feel we are adequate change makers, which starts with taking a big step back from negative self-talk and cultivating some self-compassion.

Over time, I noticed that Charlie had started to develop some self-compassion and was no longer limiting his self-image to that of a human sack of fossil fuels. This was particularly clear when he told me he was becoming open to the idea of not giving up on music. He was no longer framing it as a carbon-emitting career that he must deprive himself of in order to be comfortable in his own skin, but as something meaningful that could be flexibly enjoyed. After he moved back to Delaware, he wrote a song for the first time in almost a year, and started laying plans to set up a small home studio. He was learning to live with ambivalence—the deep understanding that none of this is perfect, and that the changes he was able to make for now were meaningful in themselves. As the world chugs towards a greener economy—nowhere near as fast as we need it to, or at a grand enough scale—our ability to get comfortable with the discomfort of our ambivalence is a crucial coping tool.

That said, there may be a healthy—and useful—way of relating to *some* feelings of guilt in this crisis. Guilt is useless when it causes us to dwell on bad feelings about ourselves, but it can be motivating when we learn to transform it with purpose. Psychiatrist Robert Jay Lifton describes two kinds of guilt: *static guilt*, where we stew in self-condemnation, and *animating guilt*, the process of "bringing oneself to life around one's guilt." When we accept our responsibility for something bad that we did, and release our guilty energy to make amends, a kind of healing can take place. This is where the spark that's needed to offer reparations can be sourced, with other people, other species, and the Earth itself.

We've established that eco-distress is a sign of health. But when Charlie was at his worst—when he could no longer get out of bed—his eco-distress had crossed over into the realm of ill health because it was disrupting his ability to function. In that state, he found himself far outside his "window of tolerance." Psychiatrist Dan Siegel coined the term *window of tolerance* to describe the optimal zone of arousal—the Goldilocks zone—in which we thrive and have full contact with the part of our brain where executive functioning happens, the prefrontal cortex. In this window, life feels like smooth sailing. But when we are hyper-aroused, we become emotionally overwhelmed and no longer fit into the window. If we aren't aroused enough, we lose our energy, go numb, get depressed, and shut down, which also pushes us below the threshold of what's required to fit into the window. The good news is that our window of tolerance can move. Trauma narrows the window, while emotional self-regulation and self-soothing can stretch it.

If we haven't had a good night's sleep in three days, we aren't eating well, and there's a lot going on, our window of tolerance shrinks, and it would be a bad time to have a sensitive conversation with someone or make important decisions. We can expand our window of tolerance, however, through resilience-building practices, such as mindfulness, gratitude journaling, self-care, paying attention to when we are at our edge, knowing when is a good time to step back, and challenging ourselves to tolerate uncertainty for longer periods of time. This is important for anyone who feels eco-anxious, because we'll be in this crisis for as long as we're alive, and these practices can help us feel joyful, relaxed, loving, and connected, even amidst terrible truths.

"Everyone needs to stretch their window of tolerance, because not only is it hard now, it is going to get harder with climate change. Without that, I don't think we are prepared to increase our capacity to bear witness to struggles, to suffering, without just feeling overwhelmed, checking out, or lashing out," climate-aware therapist Leslie

Davenport told me. So, then, when we're well outside our window, what can we do to fit back in?

Firstly, think about your own distress sensors. What are some personal indicators that show up when you are at your edge? Are you having a hard time sleeping? Eating too much or too little? Do you notice a heaviness in the chest, a clenching jaw, a knot in the stomach? Are you only able to think and talk about collapse? If you're in an overly stimulated space of high anxiety, then it is time to think about what you can do to calm your nervous system and get yourself back to baseline.

An easy initial step is to become aware of the troubling stories you're consuming, and how to balance them with other kinds of information, novels, poems, and cultural works that feel therapeutic. Charlie once told me, "Something keeping me up at night is all the news about these methane hydrates in the Arctic." He'd been obsessing about a slew of articles warning that ice-like substances in the Arctic were releasing an incredibly potent greenhouse gas, posing uncontrollable and possibly near-term catastrophe.

When we only consume bad news about the world we are headed for, we feed a condition of *pre-traumatic stress*, which psychiatrist Lise Van Susteren defines as a "before-the-fact version of classic PTSD," a diagnosis that relates to traumas of the past. As people increasingly fear that climate disruption will upend their lives, the lives of others, and parts of the world they love, Van Susteren argues that pre-traumatic stress becomes a significant burden to their mental health.

In a podcast, she retold the experience that led her to name the condition. "I would see a beautiful picture of a coral reef or the Mediterranean or some other mountains covered with trees. And rather than thinking, 'Oh my God, that's so awesome. That's so gorgeous,' and be uplifted, I would be thinking, 'Oh my God, ocean acidification with the coral reefs, and marine life dying out, and mountains being deforested or dying from beetle infestations, and forest fires.' I was the one who ended seeing all of these beautiful images that ordinarily would have evoked such a sense of euphoria and that transcendent feeling of connecting with something larger than ourselves, and instead it was the source of stress."

Pre-traumatic stress has been studied in soldiers, whose symptoms manifest as hallucinations about the future, flash-forwards, nightmares, and phobias. In a study that looked at Danish soldiers before, during, and after their deployment to Afghanistan, researchers were able to measure intrusive involuntary images of stressful future combat scenarios as signs of pre-traumatic stress. Interestingly, soldiers who experience pre-traumatic stress before deployment have been found to have an increased risk of post-traumatic stress disorder after they return from war. The hypervigilance to anything untoward that might happen seems to prime the nervous system in a sense, so it becomes more agitated when bad things do take place.

The same thing happens to many of us when we focus narrowly on the dire environmental news cycle. If you're on Instagram and Twitter all day, and all they are showing you is terrible images of ecological chaos, how does putting time restrictions on those apps, or even deleting them, make you feel down the line? Though the solution is not to turn away from the terrible truths of the world out of some groundless optimism or wishful thinking, becoming aware of how much space we give distressing news can help a lot of people. Climate-aware therapist Thomas Doherty told me, "We have to avoid black-and-white thinking. I read all of the bad news, but I also make sure that I am getting a lot of positive news in. I make sure that I read poems that inspire me while I bear witness to all of the hard stuff. It is not about avoiding negative information, it is about being big enough to take it all in." The key is to be mindful about what we're focusing our attention on.

Mindfulness, which has its roots in Buddhism and Brahmanism, is a form of metacognition: an awareness of what is happening in the mind. The awareness doesn't get rid of things in your mind, but it changes your relationship to those things. Now remember the negativity bias of the brain. It is critical that we not marinate in the negative, because it can easily take over and block out the positive. Mindfulness teaches us to notice a difficult feeling we may be having and to create some space around that awareness so that you do not reinforce it in the negatively biased brain. The primary practice is to learn to be with what is happening, skilfully, by observing it and not getting hijacked by it. It is a way to approach the suffering that shows up in the world every day with compassion, and eventually maybe even a bit of curiosity.

If you're feeling overwhelmed about the climate crisis or ecological breakdown, try grounding yourself by going to a sensory experience in your body, and find what's happening there. Feel your feet firmly on the ground, for instance, or focus on your breathing. Notice each individual sound outside your window. The simple process of moving your attention away from your thinking mind to your present physical experience has a deeply calming effect. This in turn soothes your nervous system, and can be used as a tool to move yourself back into the window of tolerance. The body can affect the mind, and the mind can affect the body, which is why this kind of mindful meditation can help promote balance and inner peace.

I had dabbled in meditation for years but only developed a meditation practice while writing this book, which I learned how to do by attending a five-day online mindfulness retreat for climate activists led by Zen Buddhist monks. The effects have been dramatically positive. I feel much calmer and better equipped to contend with my research on a day when I've been meditating than when I haven't. Even more important is how mindfulness can help people on the physical front lines of the climate crisis. American National Public

Radio reported on how veteran Californian firefighters are being brought to their knees by the scale and challenge of the fires they now have to deal with. Cal Fire, the state's Department of Forestry and Fire Protection, has fire crews doing yoga and practising mindfulness, which is helping them cope.

If you find yourself in a moment of really high emotional drama, you can also try changing your behaviour in other ways. Go to the gym, take a cold shower, call someone you love. One therapist even told me I could grab two ice cubes if I felt I was going to lose it, and just hold them as they melt. It literally cools and then calms you down. Another recommended brushing my hair, and allowing myself to feel all the sensations of the bristles against my scalp. All of these present-moment activities, if done mindfully, can ease us. For maximum benefit, try to find a park to stroll in, or take a hike in the wild. A significant body of research shows that time spent intentionally in nature for therapeutic effect—known as *forest bathing*—is linked to lower stress, restored attention, a balanced nervous system, decreased levels of anxiety and depression, and a strengthened immune system.

If you find yourself spiralling in catastrophic thoughts, see if you can name the feelings you're experiencing. "Name it to tame it" is a popular phrase in psychology, also coined by psychiatrist Dan Siegel. Rather than stirring in overwhelm, by using mindfulness we can step outside our feelings just enough to notice what they are and give them names. This kick-starts executive functioning in the prefrontal cortex and calms our limbic system. In *Emotional Resiliency in the Era of Climate Change*, Leslie Davenport suggests thinking of your feelings as guests in your home. By greeting them all and learning their names, you can host a big gathering without getting stuck in the corner all evening with just one guest. The more you practise this kind of inner curiosity and graciousness, the more rationality and perspective become available again.

On hundreds of occasions through the years, I've found myself reading a news article, scrolling Twitter, or attending a talk where I receive some shocking new bit of information about the ecological crisis, and I go into automatic doomsayer mode. The voice in my head will say, "That's insane! How can it be melting that fast? We're totally fucked. This is such garbage that we're pretending like we can prevent the worst from happening. There is still zero effective action." Now, when the catastrophizing inner monologue begins, I try to take a step back and observe it, with words like, "That's interesting that I'm feeling afraid because of the new paper on the cryosphere. I seem to be quickly moving from panic to despair, and am telling myself that action is pointless. That's a kind of hopeless feeling. I also feel anger arising because not enough has been done." Once I've approached my reaction mindfully, I feel more spaciousness between myself and my emotions, and from there I can start to scrutinize my thoughts. Do I really believe that action is pointless? Not at all. Is it really true that *nothing* has been done? Far from it.

It's becoming increasingly imperative for each of us to know what we can do to help ourselves fit back into our window of tolerance, live with ambivalence, and develop enough self-compassion to accept that the small things we can do won't fix this problem on their own and yet matter anyway. These abilities are just as important as traditionally valued attributes such as "people skills," "professional skills," and "self-discipline" for thriving in the world we've got.

Many people with strong pro-environmental values feel the urge to ignore their difficult emotions about the ecological crisis, out of fear they'll take over if they give them any space. Research shows that emotional suppression is not an uncommon coping mechanism among sustainability professionals in the UK and Canada. Climate scientists often focus on hyper-rationality in order to get on with their work and ignore what they perceive to be emotional interruptions.

But defending against anxiety this way eventually stops working, and the distress re-emerges, leaving individuals not only defenceless but with the additional burden of shame and a sense of personal inadequacy for not being able to smother it. The next stage—stress and burnout—has led several sustainability professionals to leave their field altogether.

There is an intimate psychosomatic connection between our emotions and our health, which is why these emotionally illiterate ways of coping cause more problems than they solve. Suppressing emotions drains a person, physically and mentally, putting real stress on the body. Studies suggest that bottling up our emotions can lead to a variety of poor short- and long-term health effects, affecting memory, blood pressure, and depression, to name just a few. Plus, when we avoid our emotions, we actually make them stronger. A lot of distress is caused, not by difficult feelings themselves, but by the feelings we have about our difficult feelings. We judge ourselves negatively for feeling anything "negative," which paradoxically intensifies how terrible we feel. This resistance does not need to be there, though. If we remove the judgment, and learn to create some spatial awareness and acceptance around our feelings—by naming them, observing them, and eventually learning to sit with them with an element of curiosity—they actually do move through us. And as they do, they change how we show up in the world.

The psychological tips I've mentioned—mindfulness, self-care, forest bathing, social connection, and becoming aware of and quieting your inner environmentalist critic—can help with all that. Ultimately, though, the most important thing is to let yourself experience your psychoterratic emotions rather than try to push them down or get rid of them. Not only are they vital sources of information, they're natural to feel, given what is going on. We have to learn to live with them, and to do so well means tapping into them as fuel for personal and societal transformation. As Caroline Hickman says, internal activism makes us deeper human beings. It also makes us better activists who can effect change in the world.

Unfortunately, the common knee-jerk reaction is still to ask, "Why should we turn inward now to process our emotions when we urgently need societal change?" Let's finally put that false binary to rest. When we apply emotional intelligence to our own internal world, we can more effectively come together with others and be part of pushing for the collective transformations we need. One of the other writers Charlie reached out to in the depths of his despair told him, "You know, Charlie, this is so much bigger than you. This is how change manifests itself in groups of people." Only when enough hearts break wide open will we start to heal the broken systems that are causing us to suffer this much in the first place.

If you want to get professional help for eco-distress, it is important to seek out a therapist who identifies as climate-aware. Mental health professionals are just as susceptible to denial, disavowal, and disorientation in the face of this crisis as the rest of the population. When you're struggling with eco-distress, talking to someone who minimizes the situation and ends up making you feel gaslit will only cause more harm! The quickest way to find a climate-aware therapist is to check out a resource like the Climate-Aware Mental Health Professional Directory at climatepsychology.us or climatepsychiatry.org (for North American–based therapists). The Climate Psychology Alliance in the United Kingdom and Psychologists for Future are other resources to check out. However, these organizations by no means provide exhaustive lists of climate-aware therapists. As climate-aware mental health professionals work to bring their colleagues on board, more mental health resources are emerging all the time, and in additional countries. Some of these organizations even offer free initial therapy sessions.

KEY TAKEAWAYS

- » Activism on its own is not the antidote to despair; acknowledging your feelings and connecting with others who share them, alongside taking action, is.
- » If we ignore or fail to process our difficult feelings, we risk burnout and put real stress on our bodies.
- » The aim is not to get past climate trauma, but to learn to live with it while working to reduce its harms, because it is ongoing cultural trauma.
- » We can expand our window of tolerance through resilience-building practices such as mindfulness, gratitude journaling, self-care, forest bathing, and connecting with others.
- » Consuming exclusively negative news about the climate can lead to pre-traumatic stress and spiralling in existential dread; the key is to be mindful about what we focus on, balancing good and bad.
- » Internal activism leads to more sustainable and effective external activism.

6

GOOD GRIEF

A crisis is an opportunity riding a dangerous wind.
—TRADITIONAL SAYING

Jeremie Saunders was born in Halifax, Nova Scotia in 1988 with cystic fibrosis, a genetic disease that causes abnormal lung functioning. His diagnosis came with the soul-shaking message that he probably wouldn't see the age of twenty. His whole life he's had to balance a sense of how to live with knowing that his life will be shorter than the lives of most of the people around him. At sixteen he wrote an autobiography, part of which read:

> When it comes to thinking about my future and what I want to do with my life when it comes to getting married and having kids, I don't really know what to think. About 40% of children with cystic fibrosis live beyond the age of 18. The average lifespan for those who live to

> adulthood is 30–33. If you think about it, what's the point of getting married if I'm only going to live a few years through that marriage? And having a kid would be pointless, unless I want them to grow up over half their life without a dad. These kinds of things are hard to think about and it makes me feel kind of useless.

As it turns out, he was in that fortunate 40 percent. He made it past eighteen. And even past thirty. And as he grew to accept his approaching death, his perspective on it fundamentally changed. Though his life still hangs in the balance each day, and he doesn't know when his final moment might come, he now says that cystic fibrosis is one of the best things that ever happened to him. "You see, my unavoidable fate opened up my eyes so wide that it fundamentally changed what it means to live day in, day out, moment to moment. Cystic fibrosis has essentially given me the impetus to not only accept my own mortality but to embrace my own mortality," he said to an audience at a TEDx event in Toronto in 2017. Facing death each day, he explained, creates an enormous sense of freedom. Nothing is unattainable when you might die at any moment. The idea that every minute is "bonus time" makes the world your personal playground.

Instead of letting fear consume him, Saunders uses it as a tool to strive. And despite his younger self's doubts, he did fall in love, got married, and even travelled around the world with Canadian astronaut Chris Hadfield as he trained to command the International Space Station. He also became an actor, hosted his own CBC children's television show, and started Sickboy with his best friends—a much-loved podcast about reframing illness. His use of time is a testament to his deep connection to every waking moment.

Saunders has good reason to let dread overcome him, but he is not daunted by death and wants us to know that we shouldn't be either. We could view the planetary health crisis as an invitation to reflect on our mortality in a similar way, and contemplate the way we've used our time on Earth so far. Are you happy that you spent all those late nights at the office? Used all that energy feeling hard done by? Chased money over purpose? Do you still wish you'd learned that instrument? Written that play? Cycled across the country? Learned to farm? Given more time to your kids or that activist group? When we sense our days are numbered, things become very clear. The bullshit gets brushed aside, and we home in on what matters.

Saunders is swimming with the tide of his probable young death, and it's allowed him to create an incredibly fun career and rich tapestry of relationships, the stuff a meaningful life is made of. Ever-conscious of his own nearing expiry date, he lives each day with abandon. As we contemplate the planetary health crisis as an expiry date for our current civilization, might we find a similar sense of purpose? There is, of course, a risk of searching for meaning in an overly individualistic way. The exercise becomes dangerous if it leads us to resist our collective responsibility to one another, and instead organize our time to enjoy whatever we can of private comforts and luxuries—travel, parties, great food, great sex, whatever the desires may be—in a flurry of hedonistic nihilism. And though the variables aren't opposed to one another—a joyful private life versus a life of service to others (Saunders has certainly done both)—the question remains: How can we secure death awareness as a tool for the greater collective good in times of ecological unravelling?

By grieving his future, and accepting that he can't fix his disease, Saunders has learned to sit in that tense space of uncertainty and use it to his advantage. There's a deeply creative force that lives in the space of not knowing the answer to a question like "Will I take my last breath today?" It's a springboard to doing what you value most. The key difference between what he's going through and what we're all going through is that his illness is a single man's journey, while the environmental crisis is a collective experience, and we don't know for sure where it's headed. This collective understanding changes the nature of how we can face up to what's happening, and transform the uncertainty we feel into the things we all want: well-being, creativity, meaning, purpose, beauty, laughter, love, and, ideally, a healthy and habitable planet. However, the formula is roughly the same. When we accept that we can't escape the predicament of residing on a cooking rock with one another (Mars colonization dreams aside), and grieve the losses under way with others in this same predicament, we can open up to a more present way of being in the world. However, most people don't know how to grieve for something as abstract as a sinking coastline or the idea that they might not have children or grandchildren. This type of grief requires specialized support.

Saunders says he's stopped being afraid of his decline. Accepting death while embracing uncertainty can similarly lessen our fear and dread, and make this enriched way of living more attainable. This is why dread and grief are so crucial for us now: if we finally shake off our defences and face our worst fears in our hearts, we may finally gather the energy to act intentionally in the time we still have—and prevent the worst from happening.

We have an opportunity to not only act but also dramatically reconfigure how we show up as humans at this critical moment by redefining our top priorities. The well-being of the natural world must factor into economic decisions and guide political choices if we truly want to break old patterns and band together to work for a survivable future in which humanity might even thrive. The best definition of "transformation" I've come across describes what this involves: "the ability to take new actions that are aligned with our visions and our values even under the same old pressures." We can harness historical perspective and ancestral knowledge to support us in attempting to transform, but we haphazard humans who are alive on Earth right now are the only ones who can actualize the change. Moving past denial and dealing with our emotions are the missing link for our ability to do it. The principal emotion that can help us build a better world is love, and so it is also grief.

Grief can be very tough to bear, and at times feel never-ending, but the process is invaluable because of its great transformative power. The psychiatrist Colin Murray Parkes said, "Grief is the price we pay for love, the cost of committment." It marks our entanglement with the things that matter deeply to us. The philosopher Thomas Attig tells us that grief allows us to "relearn the world" by thinking about how the world has changed when something that matters deeply to us is lost, and how that affects our relationship to places, other people, other species, and especially to ourselves. We have choices to make about who we are going to be in the face of events we can't control. Grieving is about much more than recognizing one's own feelings; it is about welcoming how those feelings can teach, change, and heal you. It strengthens our connection with the most vital things that matter in life.

Once we understand that grief is a natural consequence of love, and that we can't have one without the other, then we start to see the mutualistic power in these emotions and why we must integrate them both into our lives in order to move ahead. As Joanna Macy writes, "As your heart breaks open there will be room for the world to heal. That is what is happening as we see people honestly confronting the sorrows of our time. And it is an adaptive response."

Grief is particularly relevant to the climate crisis—a force that causes beloved hometowns to become uninhabitable after disaster, cherished landscapes to empty into barrenness, cultural traditions to dry up, and humans and other species to die. There are several theorized models of how people grieve, and these different framings on how we process loss enrich our understanding of how grieving is both meaningful and transformative. The best-known one comes from psychiatrist Elisabeth Kübler-Ross, who described the grief of the sick and dying through five stages. The first step is denial of the shocking loss. The second is anger about the new reality. The third is bargaining, where one tries to make a deal with death itself (such as "if you just give me one more week to say goodbye to my loved ones and finish this unfinished business, I'll go in peace"). The fourth step is depression, once we understand that bargaining won't work. And finally the fifth, acceptance. Acceptance brings relief and an ability to deal with whatever time is left. Many of these stages are relevant for thinking about ecological grief. For example, people often begin their journey of reckoning with the implications of climate science in a state of soft denial before connecting with a sense of depression or rage about its grave consequences. However, because the Kübler-Ross model is based on the experiences of terminally ill people—an inevitable and final process—it isn't perfectly suited to this context. We need other kinds of grief guides to get us there.

A more accurate model for how we grieve in the ecological crisis would recognize the mobilizing power of grief as more central than the finality of death. The intensity of these emotions can transform us in profound ways. Climate-aware psychologist Rosemary Randall suggests a model for grieving climate change that outlines it as a series of tasks that can be taken on or rejected. The tasks are: accepting, intellectually and emotionally, the full reality of the losses that climate breakdown causes; getting past our various forms of denial; grasping that we are fighting for our lives now (not in a generation or two, or more); working through the emotions that this truth brings up; and adjusting to the changing world around us by taking on new skills and aspects of our identity. This adjustment provides an opening for finding purpose in our pain, and the resourcefulness to take actions that will reduce more suffering and loss. Once we do that, we can reinvest back into the world the emotional energy we've spent grieving, now reconfigured as love—a necessary part of healing.

Inversely, as disasters ramp up, if we push away the opportunity to grieve for everything that is being lost, then we will remain in at least an element of denial about how much trouble we are in. The response may be to shut down one's emotions and exist in a stunted state that lacks courage and clarity, or quite possibly numb the pain maladaptively, through drugs, alcohol, manic activity, or another harmful method.

When we kick the can of grief down the road, we become disconnected and avoidant. As Randall writes, "The present continues to feel safe but at the expense of the future becoming terrifying. On the one hand, nightmare, on the other false comfort." In other words, our refusal to face difficult emotions like grief is an approach we take—consciously or not—to calm our nervous system in the present, access relative comfort, and continue to pretend that things are more or less okay. When we do this, it thwarts our ability to take the crisis seriously; we lose connection with the part of ourselves that cares enough to act, and waste more of the rapidly dwindling time we have to protect the future, which ends up actively endangering it. Rather than refusing to acknowledge the pain that is part of being alive at this time, naming the losses that are unfolding in order to let grief rise to the surface can release energy for transformative change. In the West, we drift about in a miasma of disconnection thanks to all the scrolling, binge watching, work addiction, and rampant consumerism. These distractions undergird the sky-high rates of loneliness and depression. Grief can make us come alive again, and point us towards our life's bigger purpose amidst climate breakdown, because, as the writer Elizabeth Rush puts it, "the only way to survive grief is to care for what remains with even more heart than before."

It's not easy, though. Recall from chapter 3 that ecological grief is "disenfranchised" and, even worse, often completely dismissed by others or ignored. I've run workshops on the emotional impacts of the climate crisis, and the most common feedback I've received is that people are hugely grateful to have a place where they can explore their dark feelings in a non-judgmental setting. As one participant said, "I am just so sad. It's draining, emotionally, to try to speak about my fears and concerns in my circle of loving friends and family. I feel like an imposition. I feel distant from others. Apart. Alone." The problem is, humans need to make meaning of grief communally. We are better off when we move through it with support.

In 2020, I joined a group called the Good Grief Network, which isn't a therapy group but an innovative peer support network for processing the uncertainty and grief that are part and parcel of the ecological crisis. Each Tuesday afternoon at 5 p.m. for ten weeks, I logged into a group Zoom video call with sixteen people I'd never met face to face, to share our feelings about the intensity of the times we're in and the ecological losses—both present and future—that are becoming increasingly painful. Founded by married couple LaUra Schmidt and Aimee Lewis Reau, the Good Grief Network offers a ten-step program modelled on Alcoholics Anonymous for managing eco-distress, which they run out of their home in Scottsbluff, Nebraska. They've both been through the program for adult children of alcoholics, and they know the power of sharing difficult experiences with others in a containing setting. Perhaps a similar format, they sensed, could help people cope with "negative" psychoterratic emotions that are becoming more prevalent but have no "socially acceptable" outlet.

Schmidt and Lewis Reau are often asked what their program does to take away people's pain, but they're adamant that that's not what they do. Schmidt told me, "We are inheriting a world that is already alive with the climate predicament. We're living it now and it is just going to increase. So the idea that we can get somebody into our group or into therapy and fix their eco-anxiety or climate grief is a misunderstanding of what the problem really is. A lot of people don't want to hear that. They're like, 'Oh, you're sad because the environment is not doing well.' And we're like, no, humans fucked up. We fucked up to a point that our new normal is this disorienting uncertainty. So we're not trying to fix people, we're trying to bring them to a place of living better with mass uncertainty."

The program took off after the 2016 election of Donald Trump as president of the United States, and though Schmidt and Lewis Reau regularly facilitate rounds of the ten-step program themselves, volunteers have requested Good Grief Network facilitation packages in order to run their own groups across the United States, Canada, Australia, the United Kingdom, Singapore, Austria, the Netherlands, Finland, Germany, and Japan. In addition to those countries, participants have joined groups virtually from Sweden, Norway, Italy, Denmark, Ireland, South Africa, New Zealand, and Colombia. "We teach people how to live in the moment, how to be the most alive version of themselves now, despite outcomes, and then be emotionally prepared," Lewis Reau said. Their hope is that they are planting seeds for future generations with their work, even if they won't be around to see what they grow into.

At first, I was trepidatious about participating in the program. I don't like talking about myself in groups, and am allergic to various strains of self-help language. Each meeting, however, began with a statement that made me feel I belonged: "We seek to find the delicate balance between unrealistic optimism and angry nihilism." I could get on board with that. Their non-binary approach signalled nuance and skill for coping with ambivalence—exactly what I needed more of in my life.

During the first meeting, we were introduced to the rules of operation, which boil down to: respect one another, don't try to convince anyone that their perspective is wrong, take what you want from these meetings and leave the rest if you don't agree with what another person says. Each week has a theme, which constitutes a "step," and everyone is invited at some point in the meeting to share what they are feeling in response to it.

The first step was to "accept the severity of the predicament." For an initial meeting, things got very real, very fast. Inside the four corners of my laptop screen, a bunch of strangers' heads—one non-binary person, three men, and eleven women—shared stories of environmentally linked depression, anxiety, hopelessness, and, for one person, the belief that societal collapse was coming soon and they weren't going to be alive for much longer. Normally, I'd find such a loaded discussion with strangers awkward, but surprisingly, it wasn't. Hearing how difficult other people's emotions were helped me feel less alone in my own.

The second week's step, "acknowledge that I am part of the problem and the solutions," introduced more pragmatic conversation. People spoke about their guilt around enjoying consumerist things—holidays, interior design, fast fashion. There were some eccentric solutions mentioned for lessening one's impact on the planet. A woods-dwelling woman shared her sustainability hack: learn how to shoot a .22 in order to hunt squirrels and eat their meat.

The third week's step was "practise being with uncertainty." At the time, this was the step I struggled with most. I explained to the group how, on one side, I was regularly hearing people—including some scientists—discuss the strong likelihood of near-term global systems collapse, and, on the other, prominent leaders I respected were sharing visions of an optimistic future fuelled by boundless renewable energy. I could hardly bear the whiplash. I explained how it felt like having multiple personalities, and that I was suffering from the lack of cohesion. After my three minutes of rambling were up, a woman with long silver hair said that she used to feel the same as me, and that she learned to tolerate her split sense of self by meditating (this was before I'd developed my own meditation practice). Several others added that the key is simply to stop living in the future, and to be here now. And though these were staples of spiritual and therapeutic teachings I'd heard before, I logged off feeling much more resourced than when I'd logged on. Sometimes, you just need to be reminded of the tools that are available, and working in groups can help with that.

Step four was "confront my own mortality and the mortality of all." Our homework was to watch a documentary about Terror Management Theory, which is informed by the work of anthropologist Ernest Becker. Becker argued in his Pulitzer Prize–winning book *The Denial of Death* that most human behaviour is driven by an avoidance of confronting one's own inevitable annihilation. Our awareness of our own mortality, he argues, is so profoundly unnerving that we unconsciously repress it at any cost to buffer against the discomfort it causes us. So universal is this manifestation of our death denial, Becker asserts, that it shapes all human stories and mythologies, cultures and religions, with their focus on what the afterlife entails. This, Becker explains, is why we create more evil as we believe we are fighting it. Sectarian forms of violence often boil down to people killing each other over a struggle of cultural symbolism, the only thing that can transcend one's own insignificant, fleeting time on Earth. We are deeply disquieted animals, programmed for self-preservation, and our aversion to death explains the bulk of our behaviour. We'll do almost anything to avoid the terror we feel when reminded that we're going to die.

As we discussed the documentary, a middle-aged woman shared that her fear of death is not about dying, but about not being here. She explained how she is kept awake by thoughts of the suffering that will still be here after her time is up, and it pains her that she won't be around to help. A millennial woman around my age described the deep helplessness she feels, and said, "Even if I'm not here, what will be? When you think several generations out and see that maybe there won't be any more humans, what do you do?" They spoke from a well of empathy and altruism, hoping for the world well beyond their own personal spheres, and suffering for their sense of being inconsequential to the project of protecting it.

We are a fix-it culture. An onwards-and-upwards culture. A progress-and-growth culture. The anxiety around death is precisely so intolerable because we don't know how to fix it. Similarly, we don't know how to stop the climate crisis or the degradation of nature with the limited power we hold as individuals. The challenge then becomes learning how to sit in that place of incompleteness and ambivalence, and take actions that might help anyway. Schmidt offered that when we can't fix the predicament we are in, we have to reframe it in order to bear our discomfort and learn to strategize from there. Once we give permission to difficult feelings, and learn to *be with* them, we can better see what choices we can make in a situation that we all must move on in.

The fifth step was "do inner work." Doing inner work, the internal form of activism that is explored in chapter 5, means feeling your full range of feelings, even the scary ones, without letting them consume you. It requires giving them space and, when that's done, observing that you don't need to act out maladaptively on them. Doing inner work also means recognizing that our past traumas and painful experiences live inside us and continue to affect our lives. If we don't look within to process these traumas, our external behaviour will reflect the interior troubles we haven't tended to. In this way, the work of caring for the planet and each other must start within ourselves.

On the day we met to talk about the fifth step, the coronavirus pandemic had just hit the United States, where most of us were living, and the "new normal" felt intensely disorienting. While Lewis Reau was reading the introduction, I had to turn off my webcam and weep out of anger and sadness. I was enraged that the candidate running on the strongest public health and climate change platform was not being chosen as the US Democratic Party's nominee, in the middle of a *fucking pandemic*. The climate crisis was worsening and yet was relegated to the back of people's minds as the pandemic ate up the bandwidth of global attention, and in that moment I was feeling hopeless.

When I collected myself and rejoined the Zoom, Schmidt and Lewis Reau invited us to take a pen and paper and write down everything we were feeling. A stream of jot notes poured out of me. After one minute of free writing, they interrupted us to ask us to consider this question, "What can I do right now in this moment?" and scribble whatever came to mind. Quite quickly, some agency revealed itself as I wrote a few lines about what I might have to offer. Schmidt and Lewis Reau followed this up with one more prompt: "What helps me feel grounded and calm?" By the end, my page was filled with evidence that even if we can't change the predicament itself, we can reorient our relationship to it in ways that are empowering, and therefore helpful and restorative. An emotional shift is always available, if we are mindful of how we access it.

Skipping ahead to step ten—"reinvest into meaningful efforts"—where the whole program culminated, we were prompted to reinvest the energy we'd lost from experiencing anxiety and grief into life-affirming actions that are deeply meaningful. To me, reinvestment is the continuous process of trying to live a life that makes some difference and can do some good; a life of purpose that creates beauty, connection, laughter, and love. I'd grown deeply concerned for all the young people whose futures are being compromised and I figured I might try to reinvest my energy towards helping them somehow.

Though I didn't know it at the time, this was the first step that would lead me to apply to a post-doc at Stanford University and the London School of Hygiene and Tropical Medicine, in order to research the mental health impacts of the planetary health crisis with an eye for interventions to improve young people's well-being. The program would later accept me and provide a new way to structure my life around supporting vulnerable youth, where helping them would in turn help me cope. Others spoke of reinvesting in their community groups, volunteer work, activism, self-care rituals, children, art, beauty, dancing, cooking, health, and fitness. Just by completing the ten steps, we'd already reinvested. As Lewis Reau said, "Developing and enriching our emotional intelligence is a way to take action, it just hasn't been previously viewed as such." The Good Grief Network allowed me to practise internal activism at yet another level of depth, and to have faith that the story is far from over. It equipped me with concepts I can use when I get too wound up in pessimistic thoughts, taught me that uncertainty is actually tolerable or at least something I can sit in, and pointed out the tools I have at my disposal to positively contribute to the world.

The climate crisis reminds us of existential threat in two ways, argues education scholar Cathryn van Kessel. The first is a direct reminder of death, and the second is the threat to our entire world view. To counteract unhelpful defences against the apocalyptic anxiety we feel, psychoanalyst and philosopher Robert Stolorow says we can make it more bearable by creating spaces for "emotional dwelling." This involves leaning into another person's emotional pain with your own, where you can both express your fear and darkest thoughts without anyone attempting to offer comfort, argue the validity of the feelings, or minimize their severity. It is about creating a space where you can stand in the fire, together.

Groups like the Good Grief Network provide a space for such "emotional dwelling." Group work is not just about sitting around and talking; it is about transformative *work*, and provides important exercises for building up the muscle of internal activism. It is the work that's required to cope better with what's overwhelming, and the work that can bring us to the far side of grief—where some strength and acceptance, and the courage to act, can be found.

Group work is also helpful because stories that are rich with complication and that don't have a neat ending—much like the ones often told about the planetary health crisis—tend to have a silver lining. In *Ecological Crisis, Sustainability and the Psychosocial Subject*, psychologist Matthew Adams argues that moments of uncertainty are in fact the most generative places for new stories to take shape about how we might like to live. When we're unsure of our own story, lost in not knowing what comes next, digesting this uncertainty in a group of supportive individuals can stabilize us. If they value it, co-author it, or fold parts of our story into their own lives and vice versa, we start to feel shored up. A community of listeners who are willing to hear and validate our experiences take what was once uncertain and make it coherent. All of this helps us feel more psychologically grounded, and, in turn, our well-being improves. We need other people's input to turn uncertainty into a resource.

The positive effect that programs like the Good Grief Network have in participants' lives has been studied and found to be a repeatable phenomenon. Jo Hamilton completed her PhD at the University of Reading, England, and her dissertation looked at what she calls Emotional Methodologies (EMs) for dealing with the complicated feelings the climate crisis kicks up. EMs are guides, programs, and curricula that help people express their complicated feelings rather than be at their mercy. Specifically, Hamilton looked at the ways that EMs support people to prioritize change in their lives, take action, and alleviate the burnout that is all too common in activist and climate work.

The Good Grief Network is an EM because it helps people explore and come to terms with the pain of loss related to the climate crisis, harnessing it all for a deeper purpose. There are many group-based empowerment and resilience programs, rooted in emotional reckoning, that have similarly cropped up. They each in their own way move people past denial and emotional paralysis towards deeper inner understanding of one's own responses and meaningful external actions to engage in. Various forms of nature connection, talk therapy, mindfulness, and faith-based approaches can augment this kind of support. Despite their differences, what all EMs do is generate *emotional reflexivity*, which Hamilton describes as an embodied awareness of how people respond to distress, engage with big problems, and feel about the crisis—all of which influences their sense of agency and the stories they tell about what kinds of actions matter. This is important. Without that meta-awareness, it is easy to get stuck in denial or fall into fatalism. By stretching our understanding of how we are responding, what stories we are telling ourselves, and what actions we can take, huge shifts in how we behave become available.

There is ample evidence for the capacity of EMs to help people broaden, deepen, and sustain their engagement with the climate crisis. But how? "Just by naming and enabling people to loosen up that bit about their emotions, it makes people aware, like *aha*, this is actually influencing my stories of change and my perception of my own agency. They become aware of it and it becomes something that they can reflect on," Hamilton told me.

There's something about going to those sad or scary places with others that holds weight. Coming together via an EM doesn't guarantee that participants will take any particular pro-environmental actions, and that is absolutely fundamental to the EM's power. When I told Hamilton about my own positive experience with the Good Grief Network ten-step program, she explained that the open-endedness of their tenth step—"reinvest in something meaningful"—is key. By not prescribing a specific kind of action for people to take, the program leads participants to respond in a way that feels purposeful and authentic to them, whatever that might be. That in itself is empowering, and makes whatever action they take more likely to last. This can then lead to what she calls "deep determination," which is a kind of ongoing commitment to act for environmental and social justice, "and to live the future worth fighting for in the present." EMs can, however, have significant limitations, including a lack of diversity of facilitators and participants, and a shortfall of accessibility—opportunities for future improvement.

After experiencing an EM, I feel I have deepened my connection to all that I love in this world, discovered new ways to practise emotional intelligence, and found a way to use uncertainty to strive for something more, just as Jeremie Saunders suggests. If I hadn't invested in internal work to cope better with my distress, perhaps I would not have shifted my career in order to work on this crisis beyond my efforts to describe its psychological toll with this book. A self-protecting part of me just wants to shrug my shoulders and close my eyes to what is happening, and though I can't know for sure without processing my grief, perhaps that part of myself would have been allowed to take the reins. That might have made life easier for me, but at the cost of my abandoning all that I'm attached to and care to see happen in the world.

KEY TAKEAWAYS

» Confronting death can unleash tidal waves of purpose, as uncertainty about how, when, or to what degree death might occur becomes a resource for living our most meaningful lives.

» Grieving is about much more than recognizing one's own feelings; it is about welcoming how those feelings can teach, change, and heal you.

» Models of grief for the planetary health crisis should recognize the mobilizing and transformative power of grief as more central than the inevitability of death.

- Support groups exist and are increasingly emerging to help people move through climate grief collectively.
- Emotional Methodologies help people prioritize change in their lives, take action, and alleviate burnout.

7

BALANCING HOPE AND FEAR

Critical thinking without hope is cynicism.
Hope without critical thinking is naïveté.
—MARIA POPOVA, writer

On July 14, 2021, I found myself furiously typing a post for my *Gen Dread* newsletter in a kind of trance. This was just a couple of weeks after a deadly "heat dome" heatwave smothered the American Pacific Northwest and western Canada, breaking temperature records in an unusually frightening way. Instead of merely jumping above old thresholds, it smashed them in some places by nine degrees Fahrenheit, making meteorologists' heads spin. For several days running, typically temperate Canadian cities where most homes lack air conditioning because they historically haven't needed it were hotter than Dubai. The town of Lytton, British Columbia, set Canada's all-time highest temperature record at 49.6°C (121.3°F), and the next day the town caught fire. It took only fifteen minutes from the first sign of flames until several out-of-control wildfires engulfed the place entirely, turning 90 percent of Lytton to smoke and ash.

Residents of Lytton ran for their lives and directly into several days of shock. More than eight hundred people died as a result of the heat dome in British Columbia—a toll that doesn't include those who cooked to death in nearby places such as Alberta, Washington State, and Oregon. Those who lived alone and without air conditioning were most vulnerable to the merciless heat. Journalists were filling my inbox in a flurry, asking for interviews in order to write tip-based articles on how to deal with existential dread. Their urgent questions propelled me to write the post.

At the time, this horrific event was only one of many dire warnings to have recently hit the headlines. The UN climate talks were coming up in Glasgow, and in preparation for that meeting, known as COP26, a draft of the latest Intergovernmental Panel on Climate Change report had been leaked. It warned of approaching emissions thresholds that, if exceeded, will not give our species a chance, rendering clear the difference between planetary and human survival. It read: "Life on Earth can recover from a drastic climate shift by evolving into new species and creating new ecosystems . . . humans cannot."

I've never had so many journalists reach out to me with interview requests as happened after the heat dome, and it continued well into the later summer months of 2021. The climate-caused deaths of hundreds of people and a billion marine animals in Pacific waters, as well as the rip-roar of dramatically early wildfires in some of those same western regions, triggered a new kind of terror for many who'd not previously been preoccupied with the climate crisis. The remarkable drought in California led to July blazes normally not seen until September or October. This was quickly followed by deadly flooding in Germany, China, India, and several other countries. Meanwhile, kids on TikTok were saying all of this makes them want to "unalive" themselves, and friends texted me while trying to hide from the heat, half joking about needing suicide pills. Then an article from the Associated Press reported that 2,100 drilling approvals had been given since Joe Biden took presidential office, despite his climate pledge. This was not a movie, though it felt like one. How are people supposed to keep their cool inside a system that is so casual about death and destruction?

Some of the journalists talked to me about having to rethink assumptions about their kids being able to live to certain ages they'd previously taken for granted. Others asked about their own survival and that of society. In my experience, these are not things that journalists writing for large mainstream outlets normally raise so freely in conversation. Their candidness signals how we are in the midst of a mass awakening, and along with this awakening comes a collective jolt of fear, even panic. At a pivotal moment like this, it is crucial that people respond to this deep abyss of alarm in constructive ways. As climate reporter Emily Atkin so deftly puts it in her *Heated* newsletter, "What's needed today is sustained outrage at the powerful, by those with the time and resources to express it." But stumbling out of one's daze of comfort into an existential battle for the future of humanity isn't something many people feel equipped for. Often, thoughts of helplessness, such as *What can I do? I'm only one person and this is all so baked in*, can be immobilizing, but actually, any pro-caring, pro-future, pro-environmental actions will help, and there are endless ones to choose from.

The idea of the *prospective survivor*, coined by psychiatrist Robert Jay Lifton, beautifully describes the opportunity we have to harness existential fears for something so much bigger than ourselves. While a survivor is someone who has touched death but made it out alive, a prospective survivor is someone who vividly imagines how they might perish and gets shaken to the core by the premonition's haunting effect. The journalists who asked me about their own survival and that of their children in the wake of the 2021 Pacific Northwest heat dome and widespread climate chaos unleashed that summer were clearly experiencing this, as readers of this book may be too.

Such death awareness, when harnessed, can be an engine for endurance, courage, and clarity. In his book *The Climate Swerve*, Lifton writes: "As prospective survivors we can find meaning in our actions to combat climate change. We can take on a survivor mission of preserving our habitat and embracing genuine forms of adaptation for our species. In doing so we reassert our larger human connectedness, our bond with our species." To put it simply, the prospective survivor, with a mind so sensitively attuned to the threat of complete annihilation, may hold the power to shake things up and bring about new ways of being human at this time that we sorely need.

What the prospective survivor does not do is make peace with death or collapse. She sometimes even finds joy in pushing against the seemingly inevitable, and would rather die trying, arm in arm with others just like her, than in a state of surrender. This is the kind of resilience we need now. When enough of us generate existential meaning by stepping out of isolation and into this role, we muster the life force that might actually prevent the worst outcomes from happening. It's a way of transforming fear into radical hope.

In recent years, amidst the onset of climate breakdown, the concept of hope has become unfashionable, a security blanket, all the more necessary, a dangerous distraction, the engine of action, and a sign of being out of touch or coddled, all at the same time. Hope is "such a white concept," climate justice essayist Mary Annaïse Heglar is quoted as saying in *Vice*. "You're supposed to have the courage first, then you have the action, then you have the hope. But white people put hope at the front. Their insistence on hope for all of these years has led to exactly where? Nowhere." Her words get at how hope *actually works*. We can't sit back and hope for a better future from our living rooms if we want any realistic chance of getting it. We need to get out there together and work to bring it forth.

As a Black American Southerner, Heglar carries the stories of her ancestors and of her people's ongoing struggles to advance their own well-being inside a relentless haze of historical and state-based oppression. The Black community did not close their eyes and *hope* the issues would resolve themselves, or that someone would come and save the day; they organized, advocated, and continue to push against the brutality and banality of wrongdoing that comes with a white supremacist system, continuously improving their situation. Coming at us all now is climate disruption as cultural trauma, which requires a similarly determined response rather than empty declarations of hope.

It took me a while to learn how this more robust form of hope and survival works. As noted in chapter 2, in 2011 I watched filmmakers pitch their documentary about the tragic life of orangutans in Borneo. I was disturbed by how the environmental media were glossing over reality in favour of positive emotions. Ten years later, it is hard to find any authentically positive environmental news stories, and absorbing all the negative material that is available has been tough. For a while, I couldn't stop focusing on the very worst outcomes, just as Charlie Glick did. I read books about it being too late to prevent societal collapse. I collected articles written mainly by Western middle-aged and older white men who have dominated the field, and who say we must accept that civilization is falling apart, that it cannot be stopped, and we must accept the horrors children already alive today will experience. Some spoke of extinction; others, certain bloodshed and deadly heatstroke as humans compete over dwindling resources. Like the men authoring these works, I was fascinated with our own downfall, and what kept me hooked to their words was the growing pit of fear in my stomach. I quietly disagreed with anybody who refuted these visions, brushing off their optimism as a lack of awareness of the facts, and possessed not an ounce of hope about where this was all headed. This was after my awakening in 2017, before I'd processed my emotions, and I had never been unhappier in my life.

The thing is, I hadn't yet realized that breakdown sometimes leads to breakthrough, and I was falling victim to split black-and-white thinking about what we face—that uncomfortable feeling of holding two opposing possibilities in mind at the same time.

When we're on the receiving end of overwhelming stories about the crisis, how can we avoid falling into this split thinking? Who do we listen to, and how can we interact with the darker visions while staying conscious, action oriented, and emotionally stable? Beyond just *who* you are listening to, *how many* different perspectives are you following? What mix of voices are you letting in?

It is important to question our own internal narratives, where they came from, who perpetuates them, how we can be certain of their validity, and what else might also be true. Stories such as "we're doomed" or "we're out of time" are often rooted in emotional reactivity and not evidence based. In my encounters with the "earth as hospice" crowd, for instance, I've had people tell me that they know all the following: they'll die from a climate-related civil uprising, their new grandchild will reach thirty years of age max, their adopted pet kittens won't get to live out their full lifetimes, and the entire global system will fall apart within five to ten years, taking most of the human population with it.

These claims can feel like the splattering of one's fears onto others without much consideration for their potentially traumatic effect. Several scientists have spoken out fiercely against people who claim that near-term societal collapse is inevitable, while others have supported their claims. After dealing with the aftershocks of self-confident assertions about mass death and collapse, first with horrified interest and then with scrutiny, my measured response to each of them now is: How on Earth can you *know* what is going to happen, *regardless* of what actions are taken now and into the future? Such terrible outcomes are possible if we do nothing, but it will only get as bad—and as hot—as we let it. Besides, stories of doom completely skip over how we might constructively adapt to so much disturbance. In *Emotional Resiliency in the Era of Climate Change*, Leslie Davenport writes: "The unexamined stories we repeat to ourselves are almost always distorted. There may be elements of truth, but we amplify, minimize, add layers of self-judgment, and efficiently filter out relevant and often positive aspects of the situation."

If people don't have any way of envisioning hope for the future, then they will have no motivation to act, and so stories that imply our efforts are futile become a self-fulfilling prophecy. This leads to narrative foreclosure. Psychologist Ernst Bohlmeijer and colleagues define *narrative foreclosure* as "the conviction that no new interpretations of one's past nor new commitments and experiences in one's future are possible that can substantially change one's life-story." No one wants to experience the soullessness of narrative foreclosure about their own life and the lives of everyone they love. And yet many people completely obsess about these frightening possibilities until they can see no way to avoid them. It's easy to do, because these kinds of stories are sticky—their emotional power draws us in and glues our gaze to their logic of horror, allowing us to become fixed on one specific expectation of the future without entertaining other simultaneous truths.

Such tunnel vision can have grave consequences for our well-being, which is a shame, because more nourishing combinations of stories—ones that don't gloss over how existentially challenging our situation is but still allow room for possibility and positive emergence—are also available. We can find ways to contain these multiplicities using tools like binocular vision, which I'll shortly describe. Such a shift in perspective can help us tolerate simultaneous opposing thoughts, and even be empowered by them. This is what it means to become capable of flexible thinking—an increasingly necessary lifeline for the decades ahead.

It is also wildly unjust to tightly subscribe to a vision of the future that sees our world as doomed to violent societal collapse and possibly even human extinction. This spreads the lie that we don't have any agency in our situation, other than to reach out to touch someone and tell them we love them before it all goes to shit. Those of us who are alive right now have real power over how much future suffering and loss there will be, so we don't get to give up like that—not on the wider natural world and plethora of still living species, not on young people, not on future people who aren't yet born, and not on the billions of people already living in circumstances that resemble the collapse many of us are imagining. In the words of the humanitarian designer Vinay Gupta, "What you people call collapse means living in the same conditions as the people who grow your coffee."

After being drawn for years to stories that focus on the worst outcomes, and suffering emotionally from it, I've learned that you need to approach these kinds of narratives with the right type of mental equipment. Becoming fixated on the worst outcomes can spur or strengthen depression, and outright rob you of joys you would otherwise be having. It can colour all aspects of your daily life, and sharply narrow your sense of what's possible in the world. As humans, we are creatures who need to have fun, feel alive, embrace profound connection with nature and in our relationships, as well as tap into the mysteries of the universe. It can be extremely hard to still do all of that if you're hanging out with a community that puts no stock in the future.

That's partly why, if we're not careful, the overwhelming nature of doomed environmental stories and dark eco-emotions can carry us away on a wave of fatalism, where we feel so hopeless that we close in on ourselves and the world. It's a fast-track lane to losing our finest cognitive capacities, and we cannot afford to get stuck there;

we have joys to experience in life and environmental justice to work for. To deal with this reality in a healthy way, we need to engage with it without getting crushed by it all. Crucially, we also need to feel that we can actively contribute to a collective mission of transforming the future into something better than what we currently anticipate. Our inner dimensions crave the knowledge that the actions available to us do matter.

I don't know about you, but I have never been inspired to join a cause because I felt that everything was going well. And though the central importance of "negative" emotions in environmental communications has often been dismissed, research shows that outrage, sadness, and grief can lead to increased motivation and support of environmental campaigns. Avoiding naming the dark underbelly creates the illusion that everything is fine. And while many smart people are working to improve the planetary situation, which is plenty worth celebrating even as we continue to sound the alarm, everything is not fine. Black-and-white thinking about hope and fear wrongly suggests that they aren't equally important in this quest, as well as every emotion between them.

As sociologist James Jasper writes in his article "The Emotions of Protest: Affective Reactive Emotions in and around Social Movements": "Not only are emotions part of our response to events, but they also—in the form of deep affective attachments—shape the goals of our actions. There are positive emotions and negative ones, admirable and despicable ones, public and hidden ones. Without them, there might be no social action at all." In this civilizational threat scenario, we must learn to use our emotions—the full spectrum of them—as tools.

We need to walk the fine line between the emotions that motivate us to join a movement and the emotions that nourish us as we stick with it for the long haul. They are often different. Fear might jolt you to get things done in a crisis scenario, but it can become a toxic stress if it lasts too long, causing you to tune out (hello pandemic fatigue). And if the story you're being told no longer inspires you, why not turn away? An exclusive focus on what's stressful has never been, and will never be, sustainable.

As politicians, journalists, and storytellers, we haven't done a good job of describing all the positives we'll reap by decarbonizing, and the conversation has been stuck for far too long on the environmentally harmful foods, flights, and rampant consumerism that emit so much carbon. Rather than focus on the new clean energy jobs, breathable air, and more resilient communities that a transition will create, we portray the work to be done as a giant civilizational sacrifice. In *A Field Guide to Climate Anxiety*, Sarah Jaquette Ray suggests that "reframing environmentalism as a movement of abundance, connection, and well-being may help us rethink it as a politics of *desire* rather than a politics of individual *sacrifice* and denial." Imaginings of the future are incomplete when one only sees what one wants to avoid. At the same time, we lose sight of the urgency if we don't feel the danger at all. Worrying about the worst outcomes is what we all need to do in order to spur ourselves on to take collective action that might help tip the scales. It can also motivate us to start making profound changes within our individual lives.

So it is crucial to keep an eye on the worst outcomes, because those visions feed the prospective survivor, but not to focus on them to the exclusion of everything else. What I'm getting at is a subtle but crucial balance. Being receptive to the stories that terrify us most can be motivating, on the condition that you can also make yourself big enough to take in other, more life-affirming perspectives—the stories that will mediate your fear and the hopelessness it conjures. This is the foundation that's required to see that collective trauma need not only lead to PTSD but can also be the catalyst that brings about post-traumatic growth, and a more beautiful, more compassionate world in many respects. Self-care in the planetary health crisis requires doing the work to keep your imaginative capacities alive, but that is very difficult when we're surrounded by fatalistic stories on all sides.

Imagination has always been a key survival tool with tangible effects, as well as a portal through which meaning can be found, even in the most harrowing situations. In *Man's Search for Meaning*, Viktor Frankl, a psychiatrist who survived imprisonment by the Nazis in concentration camps during World War II, writes of how he sustained himself through years of inhumane torture by means of his own imagination. Hungry and abused, he would vividly imagine the touch of his beloved wife's hand, and the look on her face should they ever be reunited. The hope that they might find each other again gave his suffering meaning, as did his dream of lecturing to audiences after the war was over about his theory of logotherapy—which articulates the search for meaning as the primary driver in people's lives, even in suffering. For that ambition to be possible, he needed the generative space of uncertainty. Though death at the hands of the Nazis seemed incredibly likely, Frankl didn't know his fate for sure. The not knowing allowed for vivid possibilities to open up, which mentally and emotionally strengthened him. He concluded that the prisoners who'd given up hope—as understandable as that was in their circumstance—were much faster to waste away.

I've suggested that we should make room for visions of our own possible death by ecological chaos or knock-on social strife in order to feed the prospective survivor within, yet I've also said we should concentrate on uncertainty, where positive imaginings of the future can grow. How can a person do both on their own? This is where the binoculars come in. With binoculars, we see that these are not contradictions, but rather are oppositions that work together for a common goal.

The psychoanalyst Wilfred Bion wrote about the concept of binocular vision as a way to understand how humans participate in group settings. For personal development, people require groups—familial, political, and communal. They're the relational environments that make us who we are. At the same time, the individual, Bion thought, is unconsciously uncomfortable with the emotional impulse of groups. Groups propel a collective of individuals to act as a cohesive unit without much self-reflection. This encourages groupthink and the tendency to gather around unifying beliefs rather than make space for people's instincts for individual expressions that could go against the norm. *Binocular vision* is a tool for observing what's really going on, allowing one to see external and internal mental forces at the same time. Comparing this ability to see simultaneous truths to what it is like to look through a microscope, Bion writes, "With one focus I see, not very clearly perhaps, but with sufficient distinctness, one picture. If I alter the focus very slightly I see another."

Bion's binocular vision for understanding how humans behave in groups is fascinating in itself, but we can leave the group part out of it in order to think about how this tool can help us balance hope and fear. An early application of binocular vision to environmental problems was made by psychoanalyst Shierry Weber Nicholsen in her book *The Love of Nature and the End of the World*. Taking the idea one step further here, I'd like you to concentrate on the Latin meaning of the word "binocular"—having two eyes. When I was spiralling in collapse narratives, worst-case scenarios, and deadly tipping points, I was reading through a single lens. My monocle didn't allow me to see oppositional forces that were also part of the picture, and this lack of perspective caused me great psychic pain. It slimmed and straightened my line of sight; and, like a telescope, it demanded that I close one eye in order to see clearly the vast blackness of what I was focusing on with the other—the outer space of climate chaos.

To protect the future to the best of our ability and enjoy ourselves along the way, we need the perspective that binocular vision affords. It allows us to maintain our state of awareness about the gravity of the situation while having the capacity at our fingertips, through a gentle change of focus, to immerse ourselves in a radically different field of view. We need these binoculars in order to be prospective survivors who are not also miserable, less creative, and unimaginative. As we honestly examine this crisis, this is the vision we require if we are going to tap into the resolve to act.

So what exactly are the nourishing perspectives these binoculars provide us with today? Soothing images of a Green New Deal going global that reverses climate change? Hardly. We sense our binocular vision in the suspicion we feel when faced with authoritative voices communicating sentiments like "as is clear to us all." Our augmented vision taps into our own deeper intelligence, which is telling us that beyond what is currently measurable, there lies a vast field of unconceived, unpredictable, and yet-to-be-articulated possibility. Binocular vision manifests in our ability to co-operate with the imagination, and be curious enough about our ignorance to see where it might lead. It is also a trust that comes from having experienced surprises many times before; we don't know everything there is to know about the mess we are in, or what can be done to adapt to it.

Climate-aware therapist Andrew Bryant sometimes sees clients who, like dogs chomping down on a bone, can't imagine anything other than horrific unravelling. To help them, he told me, "I try to validate the feeling that they're having as real and reasonable given the situation at hand . . . and say I get how scary this is and I'm scared too. But I've been scared before, and things often turn out in a way that surprises me. Not knowing can open up potentialities." On a more political note, the philosopher Isabelle Stengers said, "I want to hold on to a basic kind of ignorance: we still do not know what people are able to do. So those people who reject action by saying that 'it has already been tried,' those people are the enemy to me."

Remaining open to surprise isn't just hopeful rhetoric or an ethically imaginative orientation towards the future; it is a scientific feature of what it means to live inside complex interlocking systems. The political scientist Thomas Homer-Dixon argues in his book *Commanding Hope* that a humble form of hope, based in truth, is necessary to prevent humanity from descending into savage violence across the globe, and that we can source this hope by studying the properties of complex systems.

The climate system is a classic complex system. Climate scientists talk a lot about tipping points—a main feature of complex systems—because they exhibit what's called disproportional causation. This means that sometimes very small things can have really big effects, while really big things may not make much difference at all. If we surpass a certain threshold of melting of the West Antarctic Ice Sheet, for example, or thawing of Arctic permafrost, the whole Earth system will shift abruptly and irreversibly. Those are just a few of the climate tipping points to worry immensely about, and they're fundamentally different from what happens in non-complex systems, where small changes cause small effects and big changes cause big effects. When a small change in a complex system produces an enormous shift, that new pathway gets reinforced by positive-feedback loops, which lock in all that change. That's why tipping points are irreversible. One that flips, unleashing a cascade of environmental shifts, could be the thing that does us in—but it could also be the thing that allows us to heal some of our broken systems and better sustain ourselves.

Homer-Dixon uses the example of Greta Thunberg to bring this point to life. Back in 2017, if someone had said that a fifteen-year-old girl with Asperger's would soon sit in front of the Swedish Parliament with a sign, and that this event would catalyze a global movement of millions of young climate activists, and that this teenager would be invited to speak at the European Parliament, the United Nations, and the World Economic Forum, and have meetings with several powerful heads of state, you would have scoffed at how ridiculous that sounds. A pipe dream. But Greta's rapid ascent and the youth uprising it helped spark was a possibility, generated by our complex social system, that lay beyond the edge of our collective knowledge.

Just as a virus can suddenly appear that few people expected, and redefine nearly everything about modern life, surprises can emerge, seemingly from nowhere, and have dramatic impact, bad or good. For a shot at creating a more peaceful, sustainable future than the one barrelling at us, our psychological approach to this crisis and resolve to take action are the points that need to tip. As economist Eric Beinhocker writes,

> Humankind is in a race between two tipping points. The first is when the Earth's ecosystems and the life they contain tip into irreversible collapse due to climate change. The second is when the fight for climate action tips from being just one of many political concerns to becoming a mass social movement. The existential question is, which tipping point will we hit first?

Focusing on a singular, limited vision of what terrible things can happen in the future is a way to entrap yourself in a psychologically damaging place that's fed by anxiety or depression. Though it could motivate you to take action for a time, it won't sustain you for the long haul while things are still getting harder. Rather, it will sooner or later lead you to give up and regard the Earth as a graveyard within which we must all learn to die. In contrast, the potentialities of binocular vision are energizing and can balance out those narrowing effects.

If you're resigned to the idea that everything before us spells out a very particular vision of mass suffering, bring that thought into your arms and legs with a big inhale, and be present in the moment. Then feel that you actually *don't know* what exactly is going to happen. No one does! Congratulations, you just put on your binoculars. There's an excitement about the future that can arise in that dark place.

What's going to happen next? Sit with that question. You'll see that there is room for you to help shape the answer. It isn't that a good outcome is likely, but various good outcomes within this mess are possible, and there are real psychological benefits to tapping into the agency that is available to you to help bring them forth. We should trust uncertainty rather than fear it, by taking an honest look at how complex systems work.

However, it would be a disservice to yourself to unthinkingly trust in—or hope for—a good outcome. Again, this comes back to the dangers of expectation, and how the maxim "action is the antidote to despair" will fail us if we don't couple it with deeper internal work. As I explored in chapter 5, if our actions don't succeed in producing the outcomes we hope for, we can crash and burn out, and end up feeling much worse than if we had done nothing. An example of this can be framed through the United Nations Climate Change Conference COP15, which took place in Copenhagen in 2009 and failed to produce any meaningful agreement among the parties to reduce global emissions. It scarred the hearts of many climate activists and caused an exodus from the movement out of a sense of defeat.

Thankfully, there is another way to look at what it means to take action, even when things aren't going your way, which involves tapping into Buddhist thought. Buddhism teaches that hope and fear are mutualistic. When we hope for something, we desire a specific outcome and hope it will supplant an alternative possibility we fear might come instead. We are *grasping* for something, and our desire, which runs against the reality that we are not in control, becomes the root of our suffering. This is where mindfulness comes in again as a useful strategy. It can help us stop living as an act of expectation and open us up to the world as it is in the moment—a world we are not in control of. The ability to be present while courageous, and in touch with one's vulnerability instead of grasping for a certain outcome is the beam upon which to balance your binoculars. It's what allows us to act for the present-moment rightness of it—because it is the moral thing to do—without setting ourselves up to fall apart if things don't go as planned. We can always come back to the present moment, arriving home in what is real and not imagined, which brings a sense of calm.

A climate activist in her early twenties named Clover Hogan figured this out the hard—and helpful—way. After years of battling eco-anxiety and trying to beat it into submission, Hogan learned, with the help of climate psychologists, to lean into her painful feelings. Force of Nature, an organization she founded, engages young people to become change makers by encouraging them to welcome their eco-distress and associated feelings. In Force of Nature's theory of change, difficult emotions are vital navigational tools. A person's grief and hopelessness provide the heartbreak that reveals what matters most. When that's overlaid with their passions and unique talents, a sweet spot of available agency reveals itself, a way forward comprising the actions that are most likely to stick. As Hogan told me, "If you're into fashion, why not look at the fact that a third of the world's micro plastics come from the textile industry? If you're motivated by your gut, why not rethink that 50 percent of food is wasted in America? If you're passionate about music, why not use your art to communicate the urgency of this situation in universal language?"

Hogan is Australian. The record-breaking bushfires of 2019–20 marked the first time she grieved the loss of a part of her culture, along with a part of herself. The fires forced her to look at three billion animals that perished, and her friends who lost their homes, and think, "Okay, I really get that we are fighting for our lives, this is do or die." After several days of profound hopelessness, she realized she couldn't stay in the bottom of that U-shaped curve; she had to come up for air. She accepts that she'll descend again someday, and is now prepared for that. What Hogan finds the scariest is the thought that massively collective action may not be enough to save humanity and so many other species. But instead of banishing that terrifying idea from her consciousness, she's learned to tap into it to keep going, without clinging to expectations of what fruits her efforts will bear. At Force of Nature, she has found her purpose, and will continue plodding on because it gives her life meaning in the present moment. She explained to me how this influences her emotional state: "Now that I've gone to the dark place of grief I was afraid of, and come out the other side, I see I'm okay, and it makes me feel more authentic in my hope for the future."

KEY TAKEAWAYS

» A mass awakening rooted in collective fear is occurring as the climate crisis becomes more evident in extreme weather events that affect everyone.

» Existential resilience and robust hope can be found when we take on the mindset of the *prospective survivor*.

» Beware of black-and-white thinking about the crisis and its ability to ease the discomfort of cognitive dissonance; it amplifies inaccurate narratives about the complex system we live in and can make us either naively optimistic or inflexibly pessimistic.

» Tunnel vision focused on the worst outcomes can be broadened out with *binocular vision*, which allows us to maintain our state of awareness about the gravity of the situation and at the same time immerse ourselves in a radically different field of view.

» Imagination amidst uncertainty has always been a key survival strategy with tangible effects.

» Strength can be mined from taking action for the present-moment rightness of it, rather than because one expects a specific outcome.

PART THREE

CONNECT OUTWARD TO TRANSFORM THE WORLD

8

THE WORLD HAS ALREADY ENDED

What we pay attention to grows.
—ADRIENNE MAREE BROWN, writer and activist

After several months of being cooped up in our tiny San Francisco apartment during pandemic lockdown, where the only outdoor space was a fire escape that couldn't fit us both on it, Sebastian and I were feeling cramped. We discovered we could rent a whole house for the same price as the apartment in a canyon just one hour south of the city, which boasted mountain lions, bobcats, and migrating newts as neighbours. We moved into a modest bungalow perched over a creek in the Santa Cruz Mountains at the tail end of 2020.

As a southern Ontarian and a "Fynbo"—a person from the bucolic Danish island of Funen—we feel ridiculously lucky to be living in these mountains, which afford us the sight of redwoods from our bedroom window. There isn't a day we take the beautiful mist that rolls in over the top of the hills for granted. But on a warming planet, choosing to live here comes with serious risks.

To get to our house, you must turn down our road from a main mountain thoroughfare that connects with the highway and the nearest town. There are no other points of entry or exit—it's essentially one long bottleneck. Our road runs alongside a creek, snakes through several varieties of oak trees and sequoias, and tumbles over bridges. You'll pass more than two dozen homes before you reach our house, and several more beyond ours that zigzag up the mountainside. For all its beauty, this canyon community is not where you want to be if you need to escape from a wildfire.

Less than six months before we moved in, the San Francisco Bay Area was choked with smoke from record-setting wildfires all over the state. The gem of old growths in our mountain range, Big Basin Redwoods State Park, saw 97 percent of its acreage and nearly all its historic infrastructure burn. Farther upstate, the "August Complex fire" of 2020—at the time the largest wildfire in the state's history—was sparked by lightning strikes and spread across seven counties in the coastal range of northern California. It was dubbed a "gigafire," meaning over one million acres burned, devouring a swath of land bigger than Rhode Island. In total, that year saw over four million acres of Californian terrain go up in flames, from 9,917 independent fire incidents. More than ten thousand structures were damaged or destroyed, and thirty-three people lost their lives. For a time, San Francisco had the worst air quality in the world. Researchers found that exposure to wildfire smoke on the west coast during that time was associated with nearly 20,000 extra coronavirus infections and 750 COVID-19 deaths. And although wildfires are a normal part of life in the region, poor fire management and global warming worsening the drought were responsible for that year's unprecedented destruction.

Keenly aware of the risks, I started planning for what we'd do when the next fire season hit before we had even decided to move in. My strategy? A stocked emergency bag at the ready by the front door, and knowledge of how not to get stuck in the bottleneck thanks to a couple of sly escape routes that some locals told me about. As I write this chapter, it's still too early to test the plan; the fires are still months away and we don't know how close they'll get to the foliage that shades our canyon. In the meantime, I've tried making friends with the trees that live here. If they're going to burst into flames one day, I might as well get to know them in all their pre-scorched glory.

My daily walks and meditations have taught me to look at the trees with an appreciation of impermanence. They come and they go, often on time scales way longer than humans can meaningfully fathom. These redwoods can live up to two thousand years, which feels like forever to us yet is a blip on the geological time scale. If the study I cited in chapter 1 is right—the one that warns most trees alive today will be killed within forty years in massive die-offs, caused by heat waves and drought, if we don't quickly and dramatically curb our emissions—all the trees in this area may not stand a chance. Something else will be here instead, but we certainly don't know what.

The thought of a world with very few trees is utterly horrifying, and yet science is asking us to act immediately or else prepare for that scenario *within our lifetimes*. The anticipation of such an extreme outcome, combined with the lack of bold climate policies where they're needed most, leads many people to feel in the pit of their stomach that "it's too late." Dark thoughts ensue: *gulp, we're going extinct, we're fucked, there is nothing we can do.* The logical through line to ecological breakdown is that we are headed for an ending. The falsity of this sentiment has already been pointed out, but there are even more reasons to refute it.

The world has ended for many people, many times before. Where are the Mesopotamians? The Mayans? The Easter Islanders? It is certainly too late for them. The endings are ongoing and they occur at different scales. Thinking of what we're faced with as one big punctuated ending for our species doesn't only abandon the future, it forsakes the past. As Robert Jay Lifton writes, "Hiroshima was an 'end of the world' . . . And yet the world still exists. Precisely in this end-of-the-world quality lies both its threat and its potential wisdom." We learn from each apocalypse and sensitize ourselves to do better at preventing our worst fears going forward. Fittingly, the Greek root for "apocalypse" means to reveal, lay bare, unveil, or disclose.

Of course, it is too late to keep the biosphere as we have known it. We are going to endure further ecological losses, a hotter planet, a lot of hardship, and whatever new, intensely strained politics that will create. Rich countries and wealthy elites cannot continue to consume the way they do, and much will change—but we also can't give up. Through mass mobilization, community resilience, and political reform, we can make the losses far less devastating than they will be if we take no mitigating action at all. It is not too late to build the oases of existence we strive for, where we can practise being the kinds of people we want to be for one another, with more compassion for the Earth itself and all its life forms.

Gary Belkin, a psychiatrist who once got arrested during a protest in New York City along with a group of activists from Extinction Rebellion, told me, "When I was in jail, we did this exercise where we went around in a circle breaking the ice and team-building, and they asked what we are hopeful for, and I heard myself saying, *I am not hopeful*. We are going to see huge levels of mass destruction and it is not clear how much of human society is going to make it out of this tunnel. But what I am hopeful for is that we can build islands of what doing this whole human society thing better can look like." Though it is too late to keep the world as it is today, *it* is far from over. *It*, in fact, is the beginning of something *else*.

The emergence of something *else*, even when all that surrounds us feels toxic and oppressive, has led to many better things for people throughout history. The Nigerian-American philosopher Olúfẹ́mi O. Táíwò is a scholar of climate colonialism, and though it horrifies him to see how this phenomenon is ramping up and exacerbating deadly divides between the world's haves and have-nots, he doesn't collapse into despair. Why not? As a member of the African diaspora, he feels that his people actually have it better now than they have had it for the last five hundred years. As he told me, "Yes, tomorrow looks dire, but my ancestors dealt with the transatlantic slave trade and Jim Crow and colonial apartheid. And while what's happening today is neo-colonial and is not in any way just, it is in fact not as bad as things have been for us. And so, there is a kind of 'sky is falling' mentality that is precluded from that political perspective." Giving up just because it is hard right now is not viable; history provides reasonable assurance that doing so would forgo many better things that are yet to come.

Zooming out further still, life is by nature impermanent and in a constant state of flux. Listening to talks and podcasts about astrophysics can be very comforting when eco-distress comes knocking. The vastness of the cosmos puts the inconsequence of our civilization into sharp and calming focus. This perspective invites us to sink into the warm bath of the truth of the matter—that we are not in full control. Feeling out of control can be terrifying, but when the concept of control is reframed, healthier ways of relating to the world are revealed.

The lesson of the planetary health crisis is that when we think we are in control of nature—an anthropocentric point of view that puts humans at the centre of the universe—we quickly learn that we are not. Our self-deception has come back to bite us. Clearly, we have enough influence and agency to achieve extraordinary things, like sending research probes to Mars. But we also accidentally light the ocean on fire, as happened in 2021 when an underwater pipeline leaked in the Gulf of Mexico, causing a galaxy-shaped surge of flames, resembling a mythical sea monster's giant flaming eye, to emerge from the crest of the waves. These kinds of events show how tenuous our state of control actually is.

In *The Cybernetic Brain*, theoretical physicist turned philosopher of science Andrew Pickering highlights moments in scientific history that bring the fundamental unknowability of the world to life. His education in physics taught him that the environment we wake up to each day is a fixed and knowable place, with properties that can be studied and described. But over the years, he started to look at the history of theoretical physics, and discovered that words such as "fixed" and "knowable" are useless for describing the world as it is. When I interviewed him several years ago for a podcast, he told me, "If you look in the laboratory, a different picture appears. The world isn't fixed. It's lively and emergent, always surprising to the scientists themselves. The scientists continually react on the fly to what the world, in the shape of their machines, instruments, and experimental set-ups, does, and the world acts back, and the scientists react to that, and so on." Pickering calls this waltz between matter, people, animals, and machines "the dance of agency." Similarly, in the ecological crisis, we are confronted with the considerable agency of non-human forces, and recognizing this could unleash many wiser approaches for confronting what we are dealing with.

Since the 1970s, the field of "deep ecology"—a type of environmental philosophy that promotes the inherent value of nature rather than its instrumental worth to humans—has identified Western society's anthropocentrism as humanity's most pressing problem. Like the deep ecologists, Pickering wants us to seriously question the Western convention that says humans are the only agents in the world, and that animals, plants, and other natural beings are passive, machine-like, and predictable. He uses the example of the Deepwater Horizon disaster, another fossil fuel–caused catastophe that took place in the Gulf of Mexico, resulting in an oil spill that lasted for eighty-seven days in 2010. "Does it make sense to think that we can drill for oil a mile below the surface of the sea and control that? I think the answer has to be no. There might have been a few engineering mistakes, as one would say, but we should expect things to go wrong. The more and more desperate we become in trying to dominate the planet, the more dangerous these things become." Disasters arising from mega-engineering fossil fuel projects are piling up, and the hope is that we will draw conclusions from them about the importance of our own humility. If we question the norms many of us were taught to see nature through from the time we were children, we'll see how ill-advised it is to attempt to dominate nature at such a scale. Pickering enlarged my sense of just how emergent nature is and opened my eyes to the promise of its renewal. It is going to continue to surprise us.

Accepting human beings' lack of control over the physical world does not equate to surrender, or imply that we should not use technology, stop our activism, halt our work on ecocentric policies, or give up on our decarbonization efforts. It means we should do all those things with both eyes open, understanding that unexpected events will always occur and sometimes foil our plans. At other times, the unexpected will pleasantly surprise us. There is something deeply liberating about accepting this reality and using it to guide our decision-making. It reconnects us to the way life on this planet has always been—that is, highly vulnerable. If global power systems are ever going to slow their pathological forms of pretending—about eternal growth, about humans being nature's masters, about your favourite planet-wrecking mantra—an acknowledgement of this unpredictability is necessary. Learning to decentre ourselves is part of a regenerative mindset that starts with recognition of the substantial agency of all other species, and even of other natural features that science does not classify as living. Many Indigenous societies see life not only in plants, animals, and humans, but also in rocks and rivers.

In a BBC radio series, philosopher Timothy Morton tries to convince his listeners that "the end of the world has already happened." He argues this is because the spell that many Westerners have been under—that of human exceptionalism—has been broken by the climate crisis and the current rate of human-caused species extinction that many scientists classify as the "sixth mass extinction" event. We can finally see and feel how tenuous our species' survival is, and so our myths of being in control are quickly dissolving. I found his take refreshing—the end of the world can actually be a *good* thing. It reveals the rot at the core of the prevailing anthropocentric world view that is a major driver of this crisis, and demands that we clean it out by changing our self-image fundamentally.

Cultural historian Riane Eisler and anthropologist Douglas P. Fry describe a duality in the character of all cultures, which are either domination or partnership oriented. In the West, we are thoroughly domination oriented. Though this is not unique to Western culture, and no society is purely one or the other, a domination-oriented culture is one that ranks beings in relation to each other. According to this world view, certain men reign over other men, men reign over women, certain races over other races, humankind over animals, humankind over the rest of nature, and so on. This view implies that one becomes safe by dominating and extracting resources, which enables colonialism, capitalism, patriarchy, and racism. All this bad karma is bound to roll back, though, as the domination of resources on a finite planet will always meet hard limits that cause the system to lose function. The Hopi, an Indigenous tribe located primarily in northeastern Arizona, use the word *koyaanisqqatsi* to describe this dysfunctional state—a life that is out of balance and unravelling.

A partnership-oriented model, on the other hand, functions as an interconnected web, and values egalitarian, life-sustaining structures and mutually supportive systems. This world view, practised through time and to this day by many Indigenous societies, is closer to the way I was taught as a student of ecology—to understand the natural rhythms of the world.

The future of our species depends on which of these world views prevails. Although it may seem unlikely, the big hope is that people will choose partnership, and not just for self-preservation. As the writer Charles Eisenstein once said in an interview, "I think that the initiation that we are being offered is outside of valuing the Earth for our own survival or our own well-being, but [is encouraging us] to relate to it as a sacred being in and of itself."

It is foolish to think that we might make it through to the other side of this crisis by using the same systems of domination—of the Earth, people, species, and resources—that created the crisis in the first place. Systems change is required for sustainable survival. Complexity science has shown that the way systems are organized generates their behaviour, and the way we organize systems is based on our beliefs. Therefore, in order to change our systems, we must start by changing our beliefs.

Elizabeth Sawin is co-director of Climate Interactive, a think tank that grew out of MIT Sloan School of Management and studies relational changes in the global energy and climate systems. Plug in a certain amount of industrial activity here, and watch temperatures and sea levels rise accordingly; decrease the activity, and see the improved effects. As a subscriber to the partnership model, Sawin tries to convince people that the world is in fact already governed by interconnectivity—and partnership—between all things. To illustrate this for me, she said, "You take a bite of an apple, it wasn't you but now it is you. Where does the apple end and you begin? You think of the atmosphere as separate from yourself. But could you live longer without your little toe, which you think of as you, or the atmosphere? You are not a physical object that ends at your skin. You're more like a whirlpool or a flame in contact with everything."

Realizing that you are a whirlpool or a flame is not a trivial mental shift. Contemporary Western culture may not stand alone in its glaring alienation from nature, but many of us living in it are certainly unaware of the effects of this alienation. We have grown to become what the mythologist and psychologist Sharon Blackie calls "disenchanted." When we no longer see every bird, tree, or ladybug as worthy of our interest and engagement, we lose some of our own vibrancy as a living thing on this Earth. The prevailing view that says humans are at the top of the chain, in a superior category all our own, is emphatically unenchanted.

Where did this severance from nature in Western culture come from? Blackie and many others who've tackled this question trace the roots of our world view back to the ancient Greek philosopher Plato, who, although he believed in *anima mundi*—the soul of the world as a connection between all living things—also said that humans are the only beings capable of reason and intellect. This made them superior to all other beings in his mind. Superiority breeds disconnection, by setting one apart. This is where our loneliness, as self-regarding gods, may have got off the ground.

Plato's student Aristotle took the superiority complex further, putting Plato's ideas into a hierarchy with humans at the top, animals in the middle, and plants at the bottom of the pile. Other thinkers extended this hierarchy into the Great Chain of Being, which placed the immaterial realm on the highest pedestal, with God first, followed by the angels, and then material beings in descending order of humans, animals, plants, and finally minerals. This strengthened the idea that the material world was inferior to the immaterial realm of the heavens. The fact that humans are part of the material world led to centuries of inescapable shame around our physical bodies.

Much later came the French mathematician and philosopher René Descartes. As a young man, he had a series of dreams that led him to believe we cannot trust our senses alone to access knowledge because our perceptions can deceive us. Therefore, to learn something about the world, he argued that reason needs to be applied. As humans were the only beings capable of conscious reasoning, his views cemented our ranking as masters of the non-reasoning, non-human world. If value came from reason, then the non-human world was not only intrinsically inferior, it was also available for our use and enjoyment. His famous line "*cogito ergo sum*" ("I think, therefore I am") permeated Western culture, and as it did, it gave people permission to rank agency in the world in terms of rationality, everywhere they looked.

The Scientific Revolution was gaining ground at that time, which fundamentally transformed the way we think about nature, the world, and ourselves. The English philosopher Francis Bacon, whose thinking was influential in the late sixteenth and early seventeenth centuries, had similar thoughts to Plato, Aristotle, and Descartes, in that he understood humans to be better than all other earthly creatures. He also believed that empiricism is the only way to get at true knowledge, and popularized the idea that if something can't be measured, then it must not exist. Through this world view, the immaterial lost its sway, the scientific method was formalized, and the Enlightenment, also called the Age of Reason, was born. The reign of empiricism led to the belief that aspects of the world we could not feel with our five senses or scrutinize with our science had no value. Belief in God—an immaterial force that could not be measured—therefore presented a problem, and Humanism—the belief in human reason and human greatness—flourished.

Ironically, this was also the period when subjugation deepened within humankind itself. Eco-feminists have identified the Enlightenment as the time when male scientists dominated nature by stripping it of agency, rendering it passive, and casting it in the female gender, according to the dominant patriarchal world view. The principles of the Enlightenment therefore allowed men to devalue nature, women, and anything "other," all in the pursuit of scientific knowledge. "Scientific methods" were also used to uphold racist ideology as empirical truth.

This quick history adds up to a divided and unequal society, but empiricism does not always breed disconnection. The science of ecology has shown that species coordinate their efforts in order to maintain whole communities and ecosystem health. Symbiotic relationships have been observed within our guts, where bacteria communicate with each other using "quorum sensing," benefiting each other and their hosts. Empiricism can bring to light the interconnectedness of life; it is not the enemy. What is, is the history of using it to uphold domination-oriented ideologies as well as deny connections in the natural world that we could not measure.

That said, a person who sees humans as not only separate from nature but above it is a person who takes whatever he wants from nature. And though there are exceptions, this domination model of the world that has fed white supremacy, patriarchy, imperialism, and colonization has been the prevailing Western world order for centuries. As explored in chapter 1, it is the same world order that gave rise to the industrial age, which created our current environmental and climate crises.

And so, what the radical danger of this ecological moment calls on us to do is to muster whatever forces we can to close the chasm of separation. Stopping to notice the beauty and intelligence of creatures beyond one's small human constellation brings sincere, profound connection. Blackie describes *enchantment* as a feeling of fully participating in the world, welcoming the mystery of things that cannot be explained, and embodying a sense of belonging and connection to things outside ourselves. She writes, "To live an enchanted life is to be challenged, to be awakened, to be gripped and shaken to the core by the extraordinary which lies in the heart of the ordinary. Above all, to live an enchanted life is to fall in love with the world all over again." If we can learn to do this, even as nature is increasingly under threat, the hope and joy this instills in us can be a healthy counterbalance to the rise of dread and anxiety.

Now, with the knowledge available to many of us, there is an opportunity to move away from domination towards partnership. We can shift away from anthropocentrism, practise humility in the face of nature's emergent grace, allow ourselves to be enchanted by it, and work towards regenerating the Earth. But this is not a sweet story of overcoming learning curves a little too late. The decades ahead will see a lot of destruction and harm that would have been preventable if we had acted sooner; certain losses are baked in. As Elizabeth Sawin told me, "It isn't the way you'd want a species to learn or wake up. But if we are going to honour what has been lost and [what we'll] never get back, one way we can honour it is to wake up, at least." Partnership is the way we wake up. It can reduce the suffering that will otherwise grow, and orient us towards prioritizing life and well-being over profit. If we don't embrace it, we will continue in the direction of the last many centuries in which this crisis was created, pretending that we're fixing things even as an unspeakable darkness settles on the world.

The Great Unravelling is the name that activist and systems theorist Joanna Macy has given to the planetary predicament we're in and making worse by business as usual: species extinctions, resource depletion, and collapsing ecological and social systems. The Great Turning, on the other hand, is the name she gives to the emergence of mutualistic, life-sustaining systems that are arising and being recovered to heal the harm. These include post-growth economics, ambitious climate action from above and below, regenerative communities, eco-villages, intersectional social justice movements, re-wilding initiatives, the swells of young people rising all over the world to demand policy changes to protect their future, and more. There's no point in arguing about which of these stories—Unravelling or Turning—is "right"; they're both happening at the same time. What matters most is which one we, as individuals, choose to put our energy behind.

As author and activist adrienne maree brown so perfectly puts it, "What we pay attention to grows." What kind of a future are we focusing our thoughts on? Is it a doomed and foreshortened one, filled with carnage and competition? Or is it a more just world where solidarity can take hold, in which people understand what "enough" means when it comes to their consumption? There have been many moments in history where people have banded together against terrible odds to improve their situation and successfully achieved societal change. Why should it be any different now, just because the physics of a changing climate are hugely imposing to us?

Others have coined terms that get at the aspirational idea of partnership. Glenn Albrecht calls it the Symbiocene, which he described to me as "the period of human history where we reintegrate with the rest of life." Thomas Berry called it the Ecozoic, "a kind of Exodus out of a period when humans are devastating the planet to a period when humans will begin to live on Earth in a mutually enhancing manner." That all sounds great. Where do we sign up? And how do we choose partnership when domination may be all we have ever seen modelled at scale?

Many environmentalists like to say that Indigenous peoples—the most successful protectors of nature the world has ever known—possess the partnership skills that should guide the world now. But the lives of Indigenous peoples everywhere have been mired in theft, dispossession, and violence. Claiming Indigenous knowledge as humanity's saviour therefore carries its own risks of perpetuating harm. In the article "Indigenous Lessons about Sustainability Are

Not Just for 'All Humanity,'" Indigenous authors Kyle Whyte, Chris Caldwell, and Marie Schaefer write about the great demand on their people to supply humanity with lessons on how to live sustainably and resiliently. However, as Indigenous peoples continue to face oppression, they write that healing and protecting their own communities must be their primary concern.

Is there a way that non-Indigenous people can best support their communities' continuance while learning about Indigenous knowledge in a way that doesn't cause more damage? I consulted one of the authors, Potawatomi scholar and activist Kyle Whyte, about this. "The idea that Indigenous people have something to give to other people in terms of possible solutions and answers, I actually just think that's the wrong approach," he told me. That's because Indigenous wisdom is not knowledge one should think of as a "valuable good" that has been overlooked by today's dominant system. Any community that has survived through everything Indigenous peoples have endured must have relied on a number of different types of knowledge. It doesn't mean they've made all the best decisions, or that they know everything; rather, it implies that their survival has been supported by a portfolio of wisdom that includes ancient knowledge, intellectual traditions, as well as contemporary science. Indigenous knowledge is a story that is unfolding and ongoing, rather than something to be rebooted from the past.

For somebody who is not Indigenous and who wants to learn about Indigenous knowledge in order to become a better partner with the natural world, Whyte suggests that a constructive starting point is to ask: "What can I do to impact or benefit that story that is unfolding? What is it that my skill sets and knowledge, or things I could learn, can do to intervene in that story in a positive way?" In contrast, an unhelpful approach would be that of a non-Indigenous person who is curious about Indigenous wisdom and who tries to satisfy a sense of guilt they might have or, perhaps without fully realizing it, attempts to lift themselves up as a hero who is righting historical wrongs. Both are deeply misguided motivations for effective change. Partnership evokes allyship, and means making positive interventions in an ongoing tale. "We have a part to play too, to lift up those allies, and then whether somebody learns about Indigenous knowledge or not is just going to fall out of that work," Whyte told me.

The idea that Indigenous knowledge is something to be shared for the benefit of all humanity also skirts over an uncomfortable fact: almost anywhere you look in the world, Indigenous peoples are actively involved in a process of reconstructing, reconstituting, and recovering their traditions from the assimilating forces of colonization. Whyte compared this to trying to learn an instrument. If you're new to the piano and, every time you sit down to practise, someone asks you to perform for them, that puts you on the spot. It robs you of the time you need to get better and eventually master the art, and thus becomes a barrier to success. Similarly, Indigenous peoples are not in the business of delivering knowledge to the rest of humanity. They're in the business of recovering knowledge systems and building reliability, trust, and accountability into those systems, so they don't have to rely on ways to survive from other groups, Whyte explained. *But we just want to make the world a better place, get through this mess, and do so by centring Indigenous wisdom*, some might say. *It is a spirit of hope that brings us here. We intend no harm and do not mean to distract you from your important cultural work.* But when we insist that our intentions are good, we fail to recognize that our intentions, especially if we are privileged, are far less important than the actual impact of our actions. As Whyte explained, "Privileged people have to show that their hopefulness can be trusted by people who stand to suffer the most from climate change risks. Hopefulness can generate the ultimate bystander effect as privileged people allow for injustices to continue all the while remaining hopeful that a better world is possible."

As the ills of domination come to light, a crucial question remains: How can one genuinely partner with others in a deeply unjust society? Whyte stressed that this is the slow work of building kinship, where kinship refers to the attachments of integrity, including reciprocity, accountability, and trust. "Kinship is not 'I respect you' or 'I valorize your knowledge.' Kinship is 'Do we actually have consensuality between us, recognizing that generations of power and oppression might mean that in our lifetimes we will never achieve genuine consensuality?'" *Consensuality*, as Whyte uses the term, has both cultural and formal meanings. Culturally, it means there is a social environment that makes people feel their consent is taken seriously and that it will be respected. This is the opposite of cultural appropriation. Formally, it refers to rights and the kinds of free, prior, or informed consent that can be given on any issue, securing an architecture of agreement in which consent is not betrayed. So even if present generations can't achieve consensuality, the idea that future generations might do so if we work towards it now is well worth trying to partner over.

Partners are in the business of planting seeds for more beautiful things to grow, even if we won't be around to see what they sprout into. Now is the time to come together through an authentic wish to improve each other's well-being and a view of the long game. But there's a familiar sentiment in environmental circles that gravely threatens our ability to partner, and this needs to be discussed.

Honest question: Have you ever felt ashamed to be part of the human race? I certainly have. When I left home to study biology at university at the age of seventeen, I learned to see humans as a plague. A virus. A swarm of locusts upon the Earth. No professor ever said outright that humans were bad, but the message emanated from their ecology lectures every day; you could connect the dots with your eyes closed. Humans cause the extinction of other life forms on this planet by introducing invasive species to new lands where they threaten native species; by destroying animal habitats for agricultural development; as an effect of urban encroachment into wild spaces; and by poaching. The reality of our destructive influence, whatever form it takes, cemented the idea in my head that it would be better for the Earth if *we* just went extinct. Why would I want to orient my life towards protecting humanity, when our very nature was so repugnant?

At that young age, I hadn't yet learned the crucial subtleties of the word "we." Who exactly were these humans I was so disparaging of? Was it the people in my province struggling to feed their kids? Or Indigenous water protectors? Or Bangladeshi families whose country is so often underwater, yet whose carbon emissions are only a tiny fraction of my own? None of that made sense. My misanthropic attitude started to unravel once I learned more about the world and, specifically, systemic and historical oppression.

Our ability to protect the well-being of the people we care about doesn't only depend on us having tools that help us self-soothe on an increasingly chaotic planet. It relies on us being able to emotionally ground ourselves in ways that allow us simultaneously to envision change and to take steps towards it. Our ability to mobilize is partly shaped by our sense of injustice, and in order to get in touch with that sense, we need to feel our connection to others. Hating "all humanity" is not only a sign of a striking lack of understanding about which humans are profiting from others' suffering, it is antithetical to the very idea of partnership. What's needed is a healthy sense of injustice that, when combined with emotional regulation, helps people act on behalf of others and the environment. I was overwhelmed as a student by the injustices people inflict on the non-human world, and often I still am. But I have learned to balance out that sentiment by looking at the injustice that also runs wild *within* our species, by connecting these two struggles, and by seeing how both issues dramatically improve when a partnering mindset is applied.

So the world is ending, has always been ending, and yet is still home to many beautiful creatures, people who deserve protection, and plenty of new beginnings. If we are not an entirely mad society, the world of frequent flying, eating meat every day, and buying fresh flowers that were cut in a distant country is ending too. That was a wasteful, harmful, and disconnected world for us to live in in the first place. The question now is, what new world are we going to whip up together from here?

KEY TAKEAWAYS

» The world has been ending throughout history for many different cultural groups; the losses are ongoing and disproportional, but so too is the proof of resilience and survival.

» Giving up just because it is hard right now is not viable—history provides assurance that doing so would forgo many better things that are yet to come.

» Learning to decentre ourselves is part of a regenerative mindset that starts by our recognizing the substantial agency of all other living creatures and of natural features that science does not classify as living, although Indigenous groups have long done so.

» Moving from a domination-based to a partnership-based culture is necessary for better planetary health outcomes; Western culture must re-examine a lot of its history in order to do this. Partnership evokes allyship and should be approached thoughtfully and with care if it is not to be just another form of extraction.

9

COMMUNICATE WISELY ABOUT THE CRISIS

It may sound strange, but when climate communicators speak to ambiguity, paradox, contradictions, and uncertainties openly, we are building cultural resilience to navigate the climate crisis.
—RENEE LERTZMAN, environmental psychologist

I was once invited to give a talk about the psychological forces that interfere with how we address the climate crisis to a network of Nordic energy executives, several of whom worked in the coal, oil, and gas sectors. Thrilled by the opportunity to speak with the high-powered people most activists in the climate movement are fighting, I mulled over my approach. I didn't want it to be a fight. I wanted it to be a meeting of hearts and minds.

In my forty-five-minute talk, I discussed what I call the spectrum from denial to distress. Backed by internal industry memos, I laid hard into the outright denial that's been funded for decades by the fossil fuel industry. How often do executives in the field talk about this? I wondered. I shared psychoanalytic research about the disavowal most people peddle in and how existential terror affects our behaviour, and I outlined Sally Weintrobe's "climate bubble" concept (explored in chapter 2). I wrapped up the talk by sharing evidence about the psychological turmoil many children, quite likely the same age as these executives' kids, are now feeling about even being here.

My intention was to approach them with an open mind, to explore ideas with compassion for what they do, and not to take a moral high ground. Compassion for fossil fuel executives? At this stage in the climate crisis? Who was I kidding? But I was trying to see the complex humans behind the professional titles, and speak directly to them in a way that would lower their defences and create space for authentic emotional engagement. I felt relatively confident the talk would lead to an open, constructive discussion about how they relate to the climate crisis.

At the end of my talk, when it came time for the Q & A, the Zoom room of approximately fifty people fell silent. Not a word. Not a question. I awkwardly treaded water for several minutes, trying to lure them into sharing even the slightest reflection on what they'd heard me say. After what felt to me like an unbearable amount of time, one man said, "I think it is very emotional for us to be reminded of the terrible things that we do."

"Okay! Now we're getting somewhere," I thought. I gently encouraged him to say more, but the silence resumed.

After a few more attempts from me at stoking the dialogue, another man said, "I think I am of the generation where I will die of old age and not climate change, but that doesn't bode well for my children." Hoo boy. True! Well spotted, sir. I managed to mutter something, I'm not quite sure what, but by that point the negative affect coursing through my body was shutting down my ability to perform well. Twenty seconds later, the host abruptly adjourned the annual meeting, and we all logged off on a terribly tense note.

Later on, I received a couple of emails from attendees who explained that my talk was very intense for them to take in, and they were simply at a loss for words. As one CEO wrote: "my feelings around the subject as a whole: Armageddon kind of feeling."

I debriefed with Caroline Hickman after the talk, and asked her what I could do better next time to enable a more dynamic conversation. What she heard me asking was *What can I do to fix this?*—a rational pull to try to improve the situation and make people feel more comfortable amidst the intolerable truth of our reality. She also told me that I was experiencing emotional transference, taking these executives' emotions on as my own. What if the way I communicated didn't need to be fixed, though? she asked. "What if it is perfect and exactly what you need to be doing?" Instead of grasping for a better technique for raising difficult conversations, Hickman pointed out, I need to get better at sitting with the discomfort—holding space for the emotions—both theirs and my own.

Learning to contain the emotions that arise in conversations about the climate crisis is challenging, and it accounts at least partly for why many of us don't talk about the climate much at all. Often we assume that others simply don't want to hear about it. We think they'll get defensive or anxious or judge us for being a downer if we bring it up. Meanwhile, those other people may be holding similar assumptions about us, and this lack of mutual forthrightness perpetuates a damaging cycle of silence. What we talk about says volumes about how much something matters to us, and this in turn affects what we focus our energies on. If no one is regularly talking about the climate emergency at work or around the dinner table at home, how important can it possibly be?

Communication about the climate can be informal, between two or more people on the street or at home, and it can be professional and public, as we see with environmental campaigns. The problem is, we've been doing both types all wrong for decades.

Let's start with professional and public communication. The term "global warming," with its identification of the heating threat, proved too divisive to become the consensus term. The more emotionally neutral but all-encompassing "climate change" emerged as the choice tag line around 1988 with the first major UN climate change assessment. But as economist Eric Beinhocker points out in an essay, "People don't care about climate change; they care about life." We'll all benefit, and more people will get involved, if the stories we tell going forward reflect this.

Then there's the challenge of running educational campaigns. Environmentalists have often been tempted to throw more facts and figures at the public, and been frustrated when no change results. Rational and emotionally dead scientific language has brought credibility to the conversation while stalling behaviour change. Science communication researchers have shown time and again that "the deficit model"—the delivery of scientific information by experts to the public, with an implicit assumption that people will soak up the facts and then act in accordance with them—does not work. People process information through their own world views, beliefs, and values, which is why conservatives and liberals can look at the exact same data and go away believing different things about it.

Another common approach is for scientists, educators, leaders, and campaigners to present a moral high ground, shaming those whose behaviours and attitudes they consider unsavoury. But this tactic only attacks people's identities and raises their defences. Strictly positive and hopeful messaging also feels delusional and inauthentic, as explored in chapter 2, which prevents some people from seeing themselves in the movement. Something far more delicate and defensive is going on when we talk about this emergency. The moment demands that we honour and investigate it.

If environmental injustices are to be healed, emotional intelligence will be key to that process. It can help people find understanding, common ground, acceptance, and, importantly, the ability to offer reparations, from which forgiveness and partnership may be achieved. That's the ultimate goal here, since mutually beneficial partnership with each other, other species, and the planet is what's required to extend the time humanity has on Earth and make it worth sticking around for.

To be good partners, we must "stay with the trouble"—as the scholar Donna Haraway says—even when it is much easier to look away because the feelings that come with facing the truth of our situation are so intense. As you know, they include: anxiety, grief, shame, guilt, rage, depression, fear, worry, cynicism, pessimism, hopelessness, helplessness, overwhelm, and the sense of being torn or caught in a dilemma . . . all signatures this book has explored. But our feelings can also include radical hope, love, excitement, aspiration, justice, connection, solidarity, joy, creativity, purpose, and meaning.

Crucially, we need to find a balance in this matrix of feelings, the sweet spot where we're not just intellectually engaged with the crisis but also emotionally engaged. A place where we know how to integrate dark emotions into our lives so that when they appear, we don't fall apart. Where we're in touch with our care for the world instead of numbed by unconscious defences. And where we're able to aim for something more and see ourselves as being able to make a difference, knowing that what we do matters even if it is small. All of this requires a high degree of emotional intelligence. We have to practise it, and we need support to do so, because it is not innate to all of us. What is more often innate is becoming stressed and rigid, shutting down, projecting blame onto others, and being defensive—all the things that are common in strongman politics but will not help us get through this mess while upholding the integrity of each human life, species, and wild place. Wise communication requires that we get emotionally educated. And when we figure out how to do this in our personal lives, we can also apply this wisdom to our professional and public communications.

Some psychologists are working to give us more sophisticated tools for communicating about the crisis in ways that foster emotional awareness, and thereby don't skirt over our multi-dimensional feelings. Environmental psychologist Renee Lertzman's non-profit Project Inside Out promotes an emotionally intelligent mindset for activating change by bringing insights from clinical psychology into the realm of climate and environmental communications. Project Inside Out has called out fear, pressure, moralizing, social shaming, hitting people over the head with facts, and a strict commitment to positive storytelling as ineffective strategies for inspiring and sustaining people to take action in their lives and organizations. Instead, Project Inside Out espouses an approach that focuses on the communicator being what it calls a "guide."

The guide understands that the crisis makes people feel anxious and ambivalent at the same time as they aspire to improve the outlook for the Earth. The guide knows that unconscious defences significantly shape how people relate to what's happening and that many of us are not cognizant of this. Project Inside Out therefore encourages change makers to put this awareness at the front of their mind when they're speaking with others about the environment, in any formal or informal setting.

A key feature of the guiding mindset is that it recognizes that many of us feel caught in a "double bind" when it comes to the climate crisis and how we live our lives. As we explored in chapter 2 with regard to "the myth of apathy," it is folly to think that our life supports are unravelling simply because people don't care. Many of us care deeply about the environment and the stability of the climate, and want emissions to decline. Simultaneously, though, we feel anxious about our ability to act in the face of what feels inevitable. Part of us yearns for that flight, or a new truck, or to keep our well-paying job in an environmentally damaging profession. And so we often find ways of justifying what we "deserve" in order to keep doing as we please, which has the negative effect of dimming our care for the world. This makes us uncomfortable and ambivalent, and we suffer—often unconsciously—for it. A guiding approach helps us tune in to this vague sense of commitment, get unstuck, and free up our capacities to make necessary changes that matter.

Some counsellors use a behaviour change technique called *motivational interviewing* to help people resolve ambivalence. It was originally developed for health care, to help people quit dangerous habits such as drinking too much and smoking. Over the years, it has been adapted for non-health-care settings, and has proven to be a useful tool for shifting habits in general. Its premise is this: even though ambivalence blocks change, when you learn how to address the ambivalence in another person—and, crucially, how *not* to address it—it actually becomes the gateway to change.

The central tenet of motivational interviewing is that if you want to compel others to act, you must tame the reflex inside you that wants to promote something in the other person. As Steven Malcolm Berg-Smith, Project Inside Out's adviser on the technique, has said, "When I start to do that, it tends to activate the other side of ambivalence in the person: the voice of no change." Paradoxically, the harder we try to fix something in another person, the more we tend to activate their resistance. It's not easy to hold back from bludgeoning people with our attempts to convince, especially the ones we love, when we see change is necessary (for their health, well-being, family, or environment). Motivational interviewing lets go of all that. No longer focused on persuasion, it surrenders to the will and intelligence of the other person to discover their own capacity for change, with light guidance from an emotionally intelligent interlocutor.

To encourage mutual openness, try adopting a state of mind that makes you feel calm and curious to understand the other person's perspective. This requires deep listening; people are more likely to move in a direction of change when they feel they're being accurately understood, so empathic listening is key. Motivational interviewing requires that we operate from a place of acceptance and non-judgment, in a way that supports and acknowledges the other person's freedom and agency, and that we do so without attachment to outcome. Not being attached to outcome doesn't mean you don't care! It means that at the end of the conversation, you more or less cannot be emotionally attached to whether the person does one thing or another. If you feel yourself becoming emotionally invested in the outcome and cannot let go, remember that you are running the risk of manipulation and provoking resistance to change. In other words, you may be reactivating what Malcolm Berg-Smith calls the "righting reflex."

Instead, the aim is to use conversation to support the other person to begin exploring and talking about both sides of their ambivalence. You want to get them to name their mixed feelings out loud. Exploring ambivalence frees up energy and helps unscramble doubt as well as awaken action. Therefore, focus the conversation by drawing out good reasons for change in the other person, by asking what they think might be beneficial for them and why they want to do something. Step out of the role of telling them what to do, and in its place tease out the aspirations, insights, motivations, hopes, and dreams that are already alive in them. Above all, remember that people are more likely to adopt change if they feel they authored the idea themselves. We humans are fickle creatures.

Of course, it can be really hard to detach yourself from the outcome of whether someone changes their environmentally harmful behaviour while also truly valuing their perspective and agency to make their own decisions. Renee Lertzman affirms this, and told me that we need to learn to cultivate compassion for people's resistance.

This is clearly not an intuitive response when you are also grappling with fear and frustration about the consequences of inaction.

Suppressing the righting reflex is challenging, requiring a deliberate shift of perspective that is not natural. We can practise this with our loved ones before we try it out in the wider world. After all, we are usually quickest to pull out the hammer with the people we care most about. To help suppress that urge, Malcolm Berg-Smith uses metaphoric imagery. "I think about sitting on my hands or biting my tongue or leaving the hammer on the upper shelf," he said.

It is also crucial to carefully consider how we are communicating about the crisis across age divides. Many young people are up against real intergenerational communication barriers within their own families, which can add so much more stress to an already stressful situation. Eco-anxiety researcher Panu Pihkala told me that many children in his home country of Finland suffer with domestic tension, as their relationships with parents and grandparents who practise climate crisis disavowal or denial get seriously damaged. The same is true for many other children around the world. Sixteen-year-old American climate activist Adah Crandall spoke about her frustrations with older generations in the *New York Times*, saying, "They try to understand, but they don't . . . I am scared for my future because of the inaction of adults in the past."

Children and youth have every right to be resentful of older generations. As writer and activist Astra Taylor described in the *New York Times*, we have a "gerontocracy" on our hands. Old people run democracies and young people are getting a bad deal, as leaders continue to turn away from doing what must be done to guard children from the threats on deck. Even where kids are relatively sheltered from direct climate impacts, the limits of their ability to garner resources for themselves, vote, and work in decision-making arms of society create a condition of defencelessness. In short, what is happening to kids is unjust, and the structure of our society actively disempowers them to do much about it. Many kids know this and are using their anger about the unfairness of it all to fuel their climate activism anyway.

Kids don't deserve the degraded and dangerous world they're getting, and all adults ought to be clear about that. There is no shortage of sentimental refrains about how the current system is eating their futures. But let's revisit the idea in more detail. Firstly, it is sobering to realize just how uniquely vulnerable children are to what's happening. Global evaluations predict that kids will shoulder 88 percent of the fallout from extreme weather events, particularly those living in poverty, and that this puts a tremendous burden on their mental health. Their still-developing brains and bodies make them more susceptible to the direct impacts of natural disasters, extreme heat, and droughts, as well as the indirect effects of food shortages, migration, economic instability, and conflict.

As so many of today's youth climate activists have said, on their protest signs and before the courts, they feel terrified and betrayed. One school striker's sign put it this way: "We won't die from old age. We'll die from climate change." Another's asked: "Why Should I Study for a Future I Won't Have?" How could they not be expressing such pessimism when the World Health Organization, UNICEF, and the *Lancet* issued a report in 2020 declaring all children in the world now face existential threats from ecological degradation, the climate crisis, and predatory marketing of unhealthy and addictive products, and that no country in the world is adequately protecting children's health? How much more absurd could this situation be?

We are already seeing unprecedented rates of anxiety disorders, depression, and suicide in young people. One study found that between 2009 and 2017, rates of depression among American 18-to-25-year-olds rose 63 percent, and between 2008 and 2017 the number of them experiencing "serious psychological distress" in the last month rose by 71 percent. Though the researchers suggest this is at least partially because of unhealthy forms of digital communication that may affect mood disorders, when considered in light of the data we're now collecting on eco-anxiety, it's clear that the climate also plays a huge role in how young people are feeling.

The parent who denies that this crisis is happening or dismisses their child's feelings about it is causing real harm. But telling one's child not to worry and that it is all going to be okay isn't the solution either; that is not offering real support. Acknowledging the truth, and standing with a child in their emotions, can help them feel less alone in what is otherwise overwhelming. Accepting the truth and acknowledging culpability can help us move beyond blame into a place of supportive partnership and transformation. "We need to create more spaces where together, young and old, we can deliberate about the kinds of lives we want to live and what kinds of global future we are envisioning," child psychologist Maria Ojala writes. If we do that with children, cultivating trust as well as our capacity to imagine better futures together, perhaps something will emerge that looks like forgiveness in an unforgivable situation. As World War II veteran Paul Boese said, "Forgiveness does not change the past, but it does enlarge the future."

In order to support children in the face of the climate and wider ecological crisis, we must understand that they mirror what adults do. Child psychologist Jo McAndrews tells climate-conscious parents to imagine themselves as a Spotify playlist, and their children as a Bluetooth speaker. If stressful music is coming from the playlist, it will also be coming from the speaker. If calm music is coming from the playlist, so too will it come from the speaker. Therefore, the first thing that parents can do to help their kids with this distress is learn how to cope with their own, so that their kids will be better able to deal with their emotions.

Parents are understandably afraid of upsetting their children by talking about the climate crisis. But Caroline Hickman says that many parents would be surprised by what their kids already know.

They're on the Internet, they're talking to each other, and they're in close communion with the natural world. Insightful communication honours that they're going to know more than we expect, whether adults tell them in age-appropriate ways or it filters in through osmosis from the wider culture and experience. What parents need to do, Hickman suggests, is cultivate enough wisdom to be able to "parent in the Anthropocene." This involves knowing that your child is going to turn around in a few years' time, when they see that the Earth is becoming less inhabitable, and say, "Why didn't you tell me? Why didn't you prepare me?"

The advisable route is to tell them—prepare them—in a way that lets them know you'll be by their side. But adults should do this only after they've first processed their own difficult environmental emotions and found ways to tolerate them without reverting to denial, breaking down, or spiralling in worry. Children long to be talked to in a tone that makes them feel securely attached and deeply supported. Conversations where parents are able to say, "You're right, it is not going to be easy, but we will find a way together to manage this," are powerful opportunities for connection. When you emphasize the climate-aware and engaged communities you're building ties with, kids will know that their parents are on their team. We have no guarantee that we'll be able to protect our kids' futures, but when have parents ever had that kind of reassurance about their children's fate? The key is to shift from panicking about what is going to happen to youngsters towards focusing on what can be done to equip them with the skills and supports they need to stay afloat in turbulent waters, as we collectively steer through all this change.

It was a sad moment when one of the fossil fuel executives emailed me after my talk to share his belief that his own children don't care about the state of the environment. He wrote, "A reflection of my own life is that my kids do not care that much at all of recycling stuff, not travel with car, plane etc. They are more like my own parents regarding taking care of the planet—they hardly don't care at all :/ As my youngest son told me a couple of years ago: Dad, Mother Nature doesn't care if there are humans or not on the planet." His kid is right. Mother Nature couldn't care less if we stay or go. It is people who care whether or not we survive.

Mindful parenting in the planetary health crisis means raising kids in relationship with nature, so that they connect with the Earth, empathize with living creatures, and feel concerned about our collective future. This in turn puts the onus on parents to learn how to contain the emotions of their children when their care about the planet naturally turns into distress. The entire family project therefore requires a levelling up—namely, a wiser way of communicating about our one delicate and warming world.

KEY TAKEAWAYS

- » Containing emotions that arise in conversations about the climate takes practice and the ability to sit with other people's difficult feelings as well as your own.
- » People's widespread inability to bear and make space for these emotions perpetuates a damaging cycle of silence.
- » The "deficit model" is not an effective way of communicating about the climate, nor is shaming others from a moral high ground, or only allowing for positive storytelling.
- » Emotional intelligence and flexible thinking are desperately needed for better climate conversations and activating change.
- » Motivational interviewing demonstrates how the ambivalence that keeps one stuck in inaction can paradoxically become a gateway to change.
- » Intergenerational communication about the climate requires special care and age-appropriate honesty.
- » Parents must learn to cope with their own eco-distress in order to help their kids with theirs.

10

THE POTENCY OF PUBLIC MOURNING

The primary emotional work we need to do to deal with our inadequacy in the face of environmental destruction is the work of mourning.

—SHIERRY WEBER NICHOLSEN, psychoanalyst and writer

Standing up to face the crowd with her signature long brown braid draped over her left shoulder, Greta waited patiently, and looked around the room. She was a full head shorter than the other teenaged activists who flanked her on both sides. In front of them, a group of Indigenous youth gathered in a line, to perform an opening ceremony. They held hands, stomped to create a beat, and sang, "We've got all of our medicine, right here right now . . . Your heart needs your attention, right here right now," repeating the refrain over and over amidst other lyrics. I watched them do this through my laptop screen, an ocean away, and was struck by how the video feed cut every few seconds from their creative presentation to a close-up of Greta's face. As the Indigenous youth sang, she looked to the side. Then she looked down. Later, she looked up again. The camera loved the minutiae of Greta's humanity at least as much as the resilience-transmitting song and dance itself.

Greta Thunberg and German climate activist Luisa Neubauer had convened this press conference at COP25, the United Nations climate change meeting that took place in Madrid in 2019. The press conference was hosted by FridaysForFuture, the global student strike movement that Greta had started the year before. Though Greta and Luisa—both well-known Europeans with huge platforms—were on the bill, they decided not to speak. Instead, they shared the time slot they'd been given with youth of colour from front-line communities, whose stories of being directly affected by the climate crisis are often egregiously ignored.

A young man, Carlon Zackhras, from the Marshall Islands explained that his country is responsible for 0.00001 percent of the world's emissions that are causing rising sea levels but has only two metres of height above water to spare. Rose Whipple, a young Lakota woman from Minnesota, told her story of resisting the Line 3 pipeline. She lamented how her community is never asked what they want, and clarified that what they want is to be stood with, listened to, and centred in decisions as they're made. Hilda Flavia Nakabuye, the founder of the Ugandan chapter of FridaysForFuture, described the climate crisis as a new form of apartheid, similar to the violent racism her ancestors suffered. She spoke, with a controlled but burning anger in her voice, of how the inaction of developed countries had imprinted scars upon her unborn children and those of the whole Global South. This, she said, is why "our hearts are bleeding with fire." Three other youth also spoke, all their words heart-wrenching.

I caught up with Hilda over the phone nearly a year and a half after she spoke at COP. She was still striking from school every Friday to advocate for climate action, along with other Ugandan youth. For the first ten minutes, I listened to her describe all the ways in which her social life revolves around environmental cleanup. When she and her friends celebrate their birthdays, they plant trees in each other's names. In their spare time, they pick plastic pollution out of lakes and then use the bottles to make water tanks for people who can't afford clean water. They run campaigns against air pollution in Kampala. They host climate debates in their schools and found environmental clubs. And only when they start to feel burned out by all this do their lives sound a bit like mine was at that age. That's when they make a point of going over to each other's houses, just to hang out and relax.

Hilda calls herself a victim of the climate crisis, and says she is never not feeling its heat breathing down her neck. When she was growing up, her grandmother ran a plantation where they grew bananas, maize, beans, sweet potatoes, and cassava. But when the weather started changing—when what used to be the dry season now extended into the wet season—their crop output plummeted. First there was drought, which withered the plants. Then came heavy rains with strong winds, which battered what crops had managed to grow. Often, they'd wake up in the morning and find their whole bounty struck down. Hilda got used to seeing her grandmother burst into tears—an image that still has a great effect on her. At one point, her dad could no longer afford her tuition, and Hilda had to stay home from school for three months. "I thought a lot about what was happening during this time. My family didn't know what was going on. We kept asking ourselves, what's all this about? When I grew up, I learned that it was climate change that caused this havoc in my childhood," she said.

Today, her family can only grow some bananas on their land, but they're not fit for the market—they're infested with pests and the soils are poor. What was once a sustainable livelihood is now a small harvest just for home consumption. And over time, as the plantation was decimated along with the stability of the seasons, Hilda's dad had to find another job, in construction, while her grandmother grew too old to transition into new work.

I asked Hilda what she would like to say directly to the world's most powerful people and the citizens of high-emitting countries that are causing the most warming. "I would want them to know that the climate crisis affects every single one of us, regardless of sex or status. And that they need to grasp the demands for climate justice coming from the Global South and millions around the world, because you too will feel the heat that we feel every day. If you're committed to achieving climate justice like you say, shut down your fossil fuel sector right now and stop destroying our future. We are people, we are humans too, and we are suffering crises we never created. Make us count."

There's an energetic field that lies in the space between not counting and counting in the eyes of the powerful, and it often takes a lot of noise and upheaval to make them care. The question is, what can we do as average citizens to draw attention to these losses and usher the shift towards justice along?

Here's an example of a devastating injustice, a taboo subject, that committed citizens brought to light and made visible. In the early 1980s, a menacing virus was discovered to cause a disease that would go on to terminate tens of millions of lives around the world, including those of more than 700,000 Americans. Ronald Reagan became president of the United States in 1981, the same year the AIDS epidemic officially began. Despite the coincidence of their appearances, Reagan waited for nearly five years into his term before he uttered the word "AIDS" in public. At that point, more than five thousand people in America, predominantly gay men, African-Americans, and intravenous drug users, were known to have died from the disease.

Reagan's silence was indicative of his administration's tactic for dealing with the epidemic overall: turn the other way. For years, as the bodies piled up and no actions were taken, those living with AIDS as well as their loved ones were told they did not matter. As the AIDS memorial website reads, "There were no formal commemorations for those who had died as would typically occur in response to the loss of lives on such a mass scale." We speak unmistakably when we say nothing at all; the absence of any public recognition of those who died from AIDS suggested their lives weren't valued. Solutions cannot be devised for crises that no one in power is willing to call out as the tragedies they are.

Grief is a private undertaking that can exist solely in one's inner world, while mourning is the outward expression of grief. What we choose to mourn shows what we choose to value. Mourning carries enormous ethical and political weight. And as we see in so many ongoing struggles, by mourning publicly with others, we can stand in power against injustice. Therein lies the significance of naming names: Sandra Bland, Michael Brown, Philando Castile, Tamir Rice, Trayvon Martin, Freddie Gray, Breonna Taylor, Ahmaud Arbery, George Floyd, and far too many others. These names appear on T-shirts. On placards in bay windows. On the tongues of those who are mourning.

Thousands who died from AIDS only came to matter, in the wider public eye, through a concerted effort to change the culture. It took a painful, beautiful, and mournful fight. Ecological grief researcher Ashlee Cunsolo makes the connection between what happened then and what needs to happen now with environmental losses. She writes:

> This process would not have succeeded without the conscious creation of public acts of mourning: public testimonies and eulogies, elaborate funerals, public memorials . . . plays and theatrical productions, and even Hollywood films helped to reconstitute the AIDS body as a human body in broader social discourse, vulnerable like our own, whose suffering and destruction is tragic, grievable, and an appropriate source of mourning.

She suggests that if we map the transformative potential of mourning that was accessed by the AIDS movement, and apply it to ecological loss, we see that mass waves of death and destruction could be infused with ethical and political weight rather than disavowed. By channelling our private grief into the public work of mourning, we can render ecological losses visible, point out their root cause, and bring the tragedies—from spoiled farms to vanishing species—into the realm of mattering. Breaking silence on a large scale is what makes disappearances count.

If we learn to mourn ecological losses, would that serve beyond highlighting our own vulnerability, and lead, as Cunsolo argues, to "healing, closure, and justice"? She mentions the Canadian Truth and Reconciliation Commission's response to residential schools and the forced assimilation of Indigenous people, the Nuremberg trials, and the International Criminal Tribunals in Rwanda and the former Yugoslavia to point to the power of what happens when states publicly witness and validate other people's mourning. When private pain becomes a political working-through, possibilities for structural change emerge. Mourning is more than a movement through grief—it is a way to mobilize.

The experience of people who've mourned losses or died from AIDS cannot be superimposed onto environmental struggles. Rather, the example demonstrates the power of cultural projects to disrupt what is dominant, whether that's the dominance of straight bodies, white bodies, or non-drug-using bodies as the only mournable subjects. It offers inspiration for how we might better call attention to environmental injustices through ritual, art, and performance. Collective purpose requires noise in all its forms.

Mourning rituals contribute to healing by helping to organize our thoughts and actions in times of profound emotional disorientation. They are different from group grief gatherings like the Good Grief Network that provide a setting for processing the intimate experience of loss, in that they externalize the grief into a public and sometimes activist-oriented occasion. They also provide some structure and social support when the pillars of familiarity have been ripped out of our lives. Rituals give us ways to mourn with dignity as we express our emotions in alignment with our deepest values, regardless of what the wider culture might say about the value of our pain. Importantly, standing with community in ritual cultivates resilience by allowing rage and anger, fed by injustice, to go somewhere outward so that it will not cause inward harm. By uniting in ritual, we become empowered to bring about cultural change. And sometimes, doing so can even make us laugh.

Rituals aren't broadly or uniformly practised in Western society. It's not that no Western people practise them, but we do so in pockets of community specification that aren't celebrated widely by the dominant society. As artist and activist Persephone Pearl told me, "In the culture I inhabit, there is a kind of ritual illiteracy and a fear and awkwardness around the use of rituals. I think I could say quite accurately that the UK is massively ritually deficient. We are distanced and alienated from place-based rituals. The rituals now are birthdays and Christmas, or personal rituals around shopping, so I am really curious about the possibilities of a ritual renaissance for the Anthropocene."

Pearl helped create a ritual called Remembrance Day for Lost Species, which now happens every year on November 30. The first was held in Brighton, England, in 2011, and the ritual has since been taken up by groups all over the world. The day has often been organized around the burning of a symbolic body of a lost species that is made beforehand. People gather, make offerings, and say whatever they feel moved to say before the body goes up in smoke.

At the first Remembrance Day for Lost Species, by Brighton Pier, seagulls circled overhead as a child with a walking stick led a procession of two dozen adults and kids, all in face paint, bowler hats, or black clothes, playing French horns, banjos, drums, and bagpipes. If you happened to pass by, you might have thought they were an anarchist marching band. In the middle of the procession, a few people walking with their heads bowed were carrying a red canoe. In it sat a large papier mâché effigy of a great auk, the flightless seabird that humans drove to extinction by the mid-1800s.

When they reached the water, a man in a homemade great auk mask read aloud from his diary against a characteristically British grey sky, and yelled in a sorrowful tone, "The great auk is gone!" A woman with a large white feather sang a dirge in the centre of a group of mourners. As it started getting dark, they pushed the canoe out into the sea, but no one had rehearsed how to set fire to an effigy of a great auk on a canoe. When they tried, it immediately sank beneath the waves.

"Invariably, at every Remembrance Day for Lost Species I've been at, we've messed it up. The effigy doesn't burn, or it blows away, or it is so cold that we can't stay outside, or someone's kid is crying the entire time. Even if the ritual is sort of awkward or inadequate or makes people feel a bit uncomfortable, I kind of love that! It is totally permission giving," Pearl said.

The permission to experiment, and mess up, and do things unrehearsed is key to the entire ceremony. A procession of activists banging drums and summoning you to grieve can very easily feel pompous. Pearl is desperate to puncture that, and welcomes laughter at any point. Sometimes people will share funny stories of people they've lost, and the overall whimsicality of their processions can evoke a jester-ish feeling. Besides, she says, "We're not doing a normal thing. We're not going home and having dinner. We're having this strange time *out of time* and that opens up space for people to have deeper conversations." Besides the great auk, Pearl has given creative cremations to the passenger pigeon, Caribbean monk seal, Darwin's toad, thylacine, Steller's sea cow, and a variety of lost and disappearing pollinators.

Pearl thinks these rituals provide a way to intervene in the hyperobject of environmental unravelling that is so hard for our minds to grasp. They give people actions they can do to at least try to think about these massively disavowed realities. The languages of economics and science that dominate how we relate to environmental change, she told me, "deny people the chance to feel like an organism in an organic context." The ritualistic, artistic, participatory, fun, and surprising activities of Remembrance Day for Lost Species are critical challenges to that.

One day, after several years of doing these rituals, Pearl, who is white, had an epiphany about the pervasive whiteness of the style of environmentalism represented by Remembrance Day for Lost Species. The Brexit vote and election of Donald Trump as president of the United States awakened her to the potency of white supremacy in the world. She reflected on how her activism networks largely consisted of privileged white people fighting ecocide by addressing capitalism and consumerism, without any analysis of the racialized underpinnings of those very structures.

In a blog post titled "On Racism and Environmentalist Practice: Reflections on a Journey," Pearl and her fellow organizers write: "Rather late in life, we realized that our brand of environmentalism was a product of racial and class privilege—and worse, that its 'colour blindness' colluded in the ongoingness of white supremacy. Privilege had led us to assume it was acceptable to focus on biodiversity loss without building this work on a foundation of solidarity and anti-racist practice." They realized that Remembrance Day for Lost Species, unless re-examined, would be irrelevant for anyone who was economically marginalized, on the front lines of the climate crisis, affected by intergenerational trauma from colonialism, or facing human rights violations, as well as anyone living in kinship with endangered species and landscapes. They've since switched their lens to focus more on the interconnection between people, species, and places than on any single species, and changed their research reading list to focus on authors and artists of colour. Pearl now thinks even the name "Remembrance Day for Lost Species" misses the mark, because "lost" seems too apolitical.

The extinction of a species doesn't occur in isolation. Because any species is always ecologically embedded, tied up in relations with other creatures, what's at stake is not just the species' own existence but the role it plays in all those other lives. Thomas Van Dooren, a philosopher of extinction, writes, "Attentiveness to the relationality and interdependence of life is particularly important because the death, and subsequent absence of a whole species, unmakes these relationships on which life depends, often amplifying suffering and death for a whole host of others."

For three thousand years, a Parsi community in Mumbai has traditionally disposed of their dead in "towers of silence"—circular, raised structures built by Zoroastrians (followers of a religion based on the teachings of the Iranian prophet Zoroaster) to contain the deceased and avoid contaminating the soil with corpses. The process is accelerated by vultures and other scavengers who come to eat the flesh off the bones. But as the vultures have disappeared, these funerary practices have had to be revised, and according to many people in the community, they are no longer functional. Van Dooren has studied the impact of the decline of the vultures on the Parsi people in Mumbai, and when I interviewed him years ago about "de-extinction," the topic of my first book, he told me, "I encountered in India a completely different way of experiencing the loss of a bird than I'd ever known. Birds don't matter to me in that way because I don't depend on them for my funerary practices. Each species has very unique entanglements . . . Part of mourning is coming to recognize those entanglements and how they matter."

The renewed ethos of Remembrance Day for Lost Species is committed to honouring such entanglements. In 2019, they held their first event that did not ritualize a single species, choosing instead the theme of "original names" to highlight the ripple effects on human lives that ecological shifts often produce. The art they commissioned that year was an attempt to recognize that species are embedded in cultures and local contexts. However, Pearl feels that it wasn't nearly as emotionally compelling for participants as picking one iconic species.

The rituals that Remembrance Day for Lost Species initially created have challenged people to do something that is still, for many, not customary: mourn the non-human world. Now it is challenging people to mourn non-human–human relations and attend to the often silenced interactions between systems of oppression and ecocide. Remembrance Day for Lost Species is a form of cultural agitation. The individuals behind it are engaged in an ongoing process of bringing complex and interconnected issues such as environmental racism, inequality, and ecological loss to the forefront of people's minds. That shift is creating a more inclusive type of art and activism that can speak to broader swaths of people around the world, even if it is harder to motivate some Brits to get on board.

Ultimately, bridge building between different struggles that are all connected beneath the surface is what the environmental movement needs to invest in doing if it is to become more effective. Generations are changing, and in multicultural nations, today's young people can talk about intersectionality in their sleep. They know why this is important because they live its direct effects, and will be bringing these values with them as they eventually take over society's commander roles.

In this time of ecological unravelling, there is a need for more parallel times *out of time* that we can step into together, where it is acceptable to say, "I'm not personally okay with this. I'm feeling absolutely devastated." Places where we are spared the response that makes us feel ashamed or overly sensitive, or that tells us, "Cheer up!" When we create new rituals for doing this, we create new cultural norms and give permission for others to do so too.

If Pearl's project is an early example of the kinds of cultural agitation that might make ecological losses matter more in our collective consciousness, in a manner similar to the way those who died from AIDS came to count, what else should mourners keep in mind? As we've seen with Black Lives Matter, a movement that harnesses mourning as a political tool, it means something when losses are named out loud, repeated regularly, and never allowed to be forgotten. The names of victims of police violence should be stated in public and in private, permeating all scales of social connection. They should be commemorated—with art, vigils, marches, anything that mourners feel moved to do. When we constantly pierce the blanket of silence and commit to a slogan like "No Justice, No Peace," people in power are eventually forced to pay attention and respond to grievances. They are made to witness people's pain, and connect with their own culpability for having allowed it to continue in the first place.

Mourning ecological losses in a way that is honest about our vulnerability may be able to incite massive public participation and spark group action, beyond what the climate movement is already doing. In all these ways, getting loud and creating rituals for mourning are effective methods for not giving up on the past, nor forsaking what can still be saved.

KEY TAKEAWAYS

- » What we choose to mourn shows what we choose to value.
- » Mourning is more than a movement through grief, it is a way to mobilize.
- » Mourning rituals contribute to healing by helping to organize our thoughts and actions in times of profound emotional disorientation.

» Standing with community in ritual cultivates resilience by allowing upsetting feelings and anger to go somewhere outward so that they will not cause inward harm.
» Mourning rituals are specifically well suited to drawing out the interconnections between social and environmental injustice.

II

STRONGER COMMUNITIES FOR A BETTER FUTURE

We've got to be as clear-headed about human beings as possible, because we are still each other's only hope.

—JAMES BALDWIN, *A Rap on Race*

On September 1, 2019, Hurricane Dorian slammed onto the island of Abaco in the Bahamas with winds of 185 miles per hour, gusts of 220 miles per hour, and storm surges of twenty feet. In the wee hours of the next morning, the raging Category 5 storm expanded onto the island of Grand Bahama, pummelling both places for some forty-odd hours. Seventy thousand people were displaced and thirteen thousand homes were destroyed. Airports were completely submerged, the water table was encroached upon by salt water, and 100,000 barrels of oil spilled in East Grand Bahama from a storage terminal owned by a Norwegian oil company. A mass of splintered wood, as though strewn from an enormous box, occupied the place where homes once stood. The economic losses were estimated at $3.4 billion, about a quarter of the country's GDP. The Inter-American Development Bank said it was equivalent to the United States losing the combined economic outputs of California, Texas, and Florida. American officials described parts of the islands as resembling the fallout from a nuclear bomb strike.

I read article after article about the hurricane, trying to imagine how these beautiful islands that my stepmom, Nancy-Jane, had taken me to as a child must now feel amidst the wreckage. Nancy-Jane lived in Nassau for eight years before she got together with my dad, and she still stays in touch with several of her old friends who live there.

One husband-and-wife duo with whom Nancy-Jane remains in contact live on Abaco and lost their houses in the storm. Unlike many of Dorian's victims, they were resourced enough to have three homes. Wanting to know more about the psychological toll of the hurricane for survivors, I asked for their contact details, and reached out via email to gently ask how they were doing. The woman explained that the month immediately following the hurricane was a particularly strange time. "We felt totally disoriented, unable to make decisions, aimless and lethargic. Both of us lost weight," she wrote. After the storm, they were evacuated to Nassau, where they sheltered for six weeks, but something was pulling at them to return to their home island. Going back to Abaco had a therapeutic effect. Once they saw how much they could do to help others in their community as well as themselves, they "snapped out of it," she told me. "Of course there are down days but far more up days. Having something to do and keeping busy are key."

Rebecca Solnit explored this altruistic impulse in her book *A Paradise Built in Hell*, which celebrates what disaster can teach us about how humans come together in the aftermath of adversity to share necessities. She writes:

> Disaster doesn't sort us out by preferences; it drags us into emergencies that require we act, and act altruistically, bravely, and with initiative in order to survive or save the neighbors, no matter how we vote or what we do for a living. The positive emotions that arise in those unpromising circumstances demonstrate that social ties and meaningful work are deeply desired, readily improvised, and intensely rewarding.

That kind of generous energy is a well-documented aspect of what typically happens after disasters. But there's also a darker side, which Naomi Klein explores in her book *On Fire*. She writes:

> We learn the same lesson over and over again: In highly unequal societies, with deep injustices reliably tracing racial fault lines, disasters don't bring us all together in one fuzzy human family. They take preexisting divides and deepen them further, so the people who were already getting most screwed over before the disaster get extra doses of pain during and after.

As it turns out, both phenomena were part and parcel of what happened with Hurricane Dorian.

From my email exchange with Nancy-Jane's friend, I learned about Americares, a health-focused development organization that was providing psychosocial services to help survivors who'd been traumatized by the storm. I made several calls and eventually managed to speak with their mental health program manager, who put me in touch with Ceonn Edwards, a Bahamian psychosocial support worker living in Freeport, Grand Bahama. Edwards rode out the hurricane at home with her husband and four-month-old baby, and described the experience to me in detail over the phone.

"Our landlord made sure that we were all battened up, with shutters on the windows and everything," she said. "At the beginning, we didn't feel much but were bracing ourselves for the impact. Once it arrived in our area, the roof started rattling and the winds were really, really heavy." Then the power went out.

Their phones still had battery and an Internet connection, so if things got worse, their plan was to run to a nearby shelter. But when they checked the local news, they read that the shelter had already had its roof blown off and was flooding. After that, there was no evacuation plan.

"There were people that died right on our street corner. They died in their homes, because of the flooding," Edwards said. Luckily, their own roof stayed on, their apartment sustained a survivable amount of leakage, and the whole family made it through. The toughest, though, was yet to come.

"It wasn't until after the hurricane that I felt much. That's when things became really hard, especially having a newborn. You always hear about these donations and trailers full of food and water, but they weren't distributed, so it was a fight for food and water for a week and a half. And there was no running water. It got contaminated with dead bodies and dead animals and oil. The authorities told us that you can't allow the water to even touch your skin, couldn't even wash dishes with it. The water was only to be used for flushing toilets." That's when her sisters in Nassau organized for her family to be evacuated.

One month after the storm, Edwards started working for Americares. As part of their relief program, they brought her on to provide emotional support to other survivors. Alongside staff from the Bahamian Ministry of Health, Americares set up a makeshift shelter in the Kendal Isaacs gymnasium in Nassau, which started its operations on September 11. The initial group of 1,200 evacuees comprised roughly 40 percent Bahamian nationals and 60 percent Haitian nationals. But by the time Edwards got to the shelter in October, the remaining evacuees were 90 percent Haitian. Many of the Bahamians had been taken in by friends and family, as Edwards herself had been. Haitians have long been in demand in the Bahamian labour force, often doing the necessary jobs no one else wants to do, and they are amply discriminated against in Bahamian society. They had far fewer resources to draw on.

The ground zeroes of the storm were the Mudd and Pigeon Peas shantytowns of Abaco, where a large proportion of the Haitians had been living. The status of the residents was mixed: some had Bahamian citizenship or permanent residency, some were undocumented; of those who had work permits, their papers would have been swept away by the wind and rain. Residents weren't counted before the storm, which made it tough to do the math later to see who was missing. It was strikingly clear, however, that many did not evacuate.

"There is a sense of hopelessness, that nothing will ever be okay, from the Abaco residents," Edwards told me. "Not only did they lose everything, they lost their loved ones as well." While the official loss of life was declared at seventy, the real number is believed to be much, much higher. When the morgues filled up, bodies were piled on top of each other in refrigerated containers. Rescue dogs sniffed corpses out from under the debris, though many were buried too deep for anyone to reach.

During the time she was working in the shelter, Edwards counselled adults, ran creative coping activities with children, and held separate group talk sessions for men and women, always ever so careful not to let it seem that she was doing therapy. "There is a stigma around mental health here. No one would ever come up to you and say, 'Hey, I need to talk, I'm feeling anxious or maybe like I have PTSD.' They would *never* do that. So I had to make them feel that I was just having a casual conversation with them." No one knew she was a counselling therapist, and it was vital that it stay that way.

One man she regularly checked up on was clearly suffering from PTSD, having intrusive, upsetting flashbacks and nightmares, and struggling to get any respite. During the hurricane, he said he ran into a church in Abaco with his wife's brother for protection. Many people were already inside doing the same. Then the whole thing collapsed. When he rolled over, he saw his brother-in-law's head smashed in. He said he was one of few survivors in the church, and was haunted by the horrors he'd witnessed.

I spoke with Edwards six months after Dorian. By then, the shelter in Nassau had closed and the building had returned to its normal function as a sports complex.

"So where did all the people go who were living in the shelter?" I asked.

"They put the residents who are families, who are also documented, into a hotel. The singles, they sent them to a school auditorium, which was only a handful of people. Anyone who was undocumented was told to find a friend or go back to Haiti, and told they'd be deported if they didn't do that. Some went back to Abaco, but in Abaco they took a risk, because if they find you, they will deport you. A lot of people took that chance and a lot got deported. Some are living in tents now, some are working. It depends."

The International Organization for Migration was tracking the return of Haitians to Haiti by the Bahamian government after the storm, and spoke with returnees who "reported they were mostly apprehended on the streets, in their place of employment, or while in their homes during raids usually carried out in the middle of the night by immigration officials. The returning migrants also indicated that they remained in detention for 10 to 30 days." For undocumented Haitians, the trauma of the hurricane, which had already hit their community the hardest, unleashed a wave of much more harm.

Edwards worked in the shelter from October 2019 to February 2020 along with one other Americares-employed mental health care specialist named Bethuel Isoe Nyachienga. Nyachienga has a long history of providing emotional support in disaster zones. He worked in Liberia during the Ebola outbreak; introduced refugee mental health care for Somalian, Congolese, Rwandan, Ethiopian, Ugandan, and Sudanese refugees in a camp called Dadaab in Kenya; and helped run a chain-free initiative in Somalia, where "mental patients" were otherwise literally held in chains. After Edwards ceased her work with Americares, Nyachienga was paired with a psychiatric nurse, and they became responsible for providing mental health services at outreach clinics in Marsh Harbour, Coopers Town, Standpoint, Fox Town, and a variety of schools and new settlements for Haitians. Half a year after the hurricane, he told me that he had helped roughly 4,000 people: 3,000 in Abaco and 1,000 at the shelter in Nassau, some of whom he saw more than once. "We are only two providing counselling services and we are not coping. One client can wear you down. We are not enough and the people who require our services are many," he said.

"What is your goal for helping people who lost everything?" I asked him.

"I'm trying to get people to a place of 'Yes it happened and we must move on' by providing them emotional support. Survivors need to accept what happened and make decisions from there. Decisions like, do I live in a single room with my child? People need to sit down and plan their own way."

I asked him if he could share a specific example of someone he'd helped.

"One guy's house was shaking so much during the hurricane that the water came in and took his wife and son. The house was demolished and they both died. He says his life now feels meaningless. He was talking about suicidal ideation. At times he cannot eat. So we've been talking to him, meeting with him regularly, and now he has been given some construction work, so his life is starting to restabilize."

When Dorian struck, it was the third Category 5 storm to pound the Bahamas in four years. As global temperatures rise, more heat energy becomes available, which increases the likelihood of tropical hurricanes in the North Atlantic and Northeast Pacific Oceans, typhoons in the Northwest Pacific Ocean, and cyclones in the South Pacific and Indian Oceans. The warmer the air is, the more moisture it can hold, and this leads to more rainfall. So as the world gets hotter, we can expect a continual boost in these kinds of intense tropical storms.

We can also expect more severe wildfires, droughts, heat waves, and floods in many parts of the world. And over the coming decades, as soils become increasingly desertified, it is estimated that hundreds of millions to well over a billion people will migrate in search of a better life. The mass movement of hungry people can have seriously harmful ripple effects. Competition between existing residents and new arrivals can flare racial and ethnic tensions, amplifying political instability and stoking conflict. Women migrating under the threat of sexual violence are known to be prone to anxiety, depression, and suicide. Farmers who remain and try to harvest their diminishing crop will often take on greater loans and may commit suicide when the pressure becomes too much to bear. Resources that communities depend on, like fisheries, will dry up and the jobs that relied on those resources will vanish. At that point, the people with the most money, education, and social supports will leave those regions to look for better prospects, leaving behind groups of less resilient people in increasingly insecure places.

As this pattern develops, the most vulnerable people will end up clustering in the most vulnerable places. Inequality will widen, and mental disorders will spread. When the next disaster violently upends people's lives, emotional aftershocks such as PTSD, suicidality, depression, domestic abuse, and substance abuse will all spike, getting worse before they get better. Poor people's property will be significantly more damaged than that of the rich. Anyone who didn't get adequately warned or given enough time to evacuate, or who had their power go down, or experienced the death of loved ones, or lost precious property, will face a much higher risk of mental illness than those who were luckier.

People who need life-saving medications will find their existence hanging in the balance of disrupted supply chains. In heat waves, people who live alone, are sick or bedridden, have no social contacts nearby, or don't have an air conditioner, or those whose neighbourhood street life deters them from leaving the house, will be the most likely to die. Heat waves will become more frequent and unbearable, and when they occur, mentally ill people who rely on antipsychotic medicines will be put at increased risk. Some antipsychotic drugs impair the body's ability to regulate its own temperature, and at sweltering degrees, these patients may experience fatal overheating. Murders, assaults, suicides, and hospitalizations for self-harm will also soar amongst the general population on especially hot days, as extreme heat makes people more aggressive. Indeed, this is all already happening.

By mid-century, humanity will find itself in one of two very different situations. Let's look at each of them in turn.

If we continue to treat mental health with the narrow, individualistic biomedical model that dominates now, huge swaths of the population will be left unsupported. One-on-one therapy sessions and hospitalizations simply cannot scale up to meet the scope of psychic damage that the planetary health emergency is causing. In low- and middle-income countries, where people are already bearing the brunt of climate trauma, mental health supports may hardly exist. In rich regions of the world that are most responsible for emissions, individual interventions may not be enough to help even the privileged few, and these formerly relatively comfortable nations will see their citizens increasingly breaking down.

Poorer regions that lack any coordinated mental health care system could get stuck in a relentless layering of life-threatening scenarios, fuelled by a downward spiral of diminishing support as health infrastructure gets weakened with each disaster. As Nyachienga said, just one patient can wear you down, and when you have thousands to look after, the resilience of the mental health care system quickly crumbles. Aid worker burnout is a serious threat to climate adaptation that begs us to come up with other models for scaling and spreading psychosocial support. The potential of intersecting crises complicates this further. Just think, how did your country react under COVID-19? How would the hospitals have coped if right as the pandemic upended everything, there were also Dorian-style hurricanes, wildfires, or floods in your region?

As the global outlook becomes more precarious, voters may turn to authoritarian leaders, who in their tough-guy mentality provide simple answers to increasingly complex problems. Facile explanations for perplexing crises help people pretend their anxiety away and adopt small-minded, self-protective rationales like "It's the immigrants' fault." Thinking they're protecting themselves, they may find this approach backfires. Countries with right-leaning governments tend to have higher suicide rates than countries with left-leaning governments. Helen Berry, a psychiatric epidemiologist who has extensively studied the mental health impacts of climate change, offered an explanation when I interviewed her. "Right-leaning ways of thinking say, 'I'm fine. If you're not fine, that's your problem. Go solve it.' Those are very distressing messages . . . and I don't think we can take that message forward and expect that there will not be an impact on the mental health and well-being of the world's population. There are huge numbers of people who are not in a circumstance where they can afford not to care." If we carry on treating mental health as a "nice to have" on top of physical health, and continue to marginalize it, separating it from the heart of what's needed to make a society function, we will be further immobilized as the climate crisis accelerates. Mental health and well-being will be sharply compromised in many places and completely disappear in others.

But there is a second vision of the future. If we transfer the bulk of responsibility for mental health care from the traditional expert-oriented, one-on-one model to a more community-based practice, the world in 2050 could look much brighter. Right now, in regions of the world with few or no mental health care specialists, lay people have been trained to provide psychosocial support. Getting outside clinical spaces and working where people live has proven to be incredibly transformative, useful, and even more effective than traditional mental health care in many cases. Vikram Patel, a psychiatrist at Harvard Medical School, is a leading expert in this "task-shifting" mental health practice, and has trained people on the ground in Goa, India, to provide counselling services to residents in their community, even though they didn't have any psychological expertise.

These lay counsellors are trained to care for people's depression and anxiety. In one clinical trial that Patel led in Goa, 70 percent of the people they helped recovered, while primary health care providers only achieved a 50 percent recovery rate with their patients. He says that psychosocial support skills can be readily passed on to non-experts by stripping away jargon, unpacking the treatment, delivering it where people are, using whoever is willing and available to deliver care, and reallocating specialists to supervise lay counsellors. This approach helps protect the human right to good mental health in regions that don't have adequate support. And it isn't only beneficial for poverty-stricken areas. If it were used in wealthy nations alongside the biomedical model, money could be saved and care more effectively spread around.

A better 2050 looks like one in which we've built up strong social connectedness and participation in our communities by starting to design for it now. At this moment when more people live alone than ever before in human history, loneliness has emerged as a serious public health problem. And at a time when our personal technologies foster dopamine-laced experiences that reward narcissistic tendencies, loud voices can seem to matter more than conscientious relationship building. Rather meaningless interactions have come to occupy a gigantic footprint in people's digital worlds. What was first sold to us under the banner of fostering connection has created a culture of constant distraction and dangerous misinformation. We need to strengthen our community ties with intention, both in real life and online. If communities are empowered to come together and develop solutions to the unique problems they face, several climate-aware psychiatrists tell me that a fair number of the mental disorders we expect to see as the crisis worsens may be prevented, better cared for, and recovered from.

The potential of social relationships to allow residents in a community to coordinate their efforts and achieve shared goals is what's known as *social capital*. People with high social capital and connectedness are less likely to develop mental health disorders in the aftermath of a disaster. One of Helen Berry's studies looked at PTSD in children who had survived a Category 5 storm in Queensland, Australia. Many of these kids had seen flying debris, broken windows, and damage to parts of their home, and in several cases they'd witnessed the entire roof coming off. The study found that the most highly connected 10 percent of the kids, defined by their having the greatest number of other people in their life to talk to, play with, and trust, were significantly less likely to have PTSD than their peers. The bottom 10 percent, meanwhile, representing the most socially isolated and loneliest kids, had significantly more and substantially worse PTSD than all the other children. This could be because social connectedness reduces feelings of loneliness, abandonment, and social isolation, all of which are states that threaten resilience.

In order for residents to protect themselves from the things they worry most about, they need governments and institutions to support them as they identify priority areas for reducing environmental harms and collectively build solutions in the places where they live. For example, a community could unite around cultivating local food security, or cleaning up toxic waste pollution, or getting residents equipped with information on how to protect themselves during the next heat wave; all of this contributes to a healthy sense of ownership of our own well-being.

Berry calls this approach, which taps into community cohesiveness, the "pearl in the oyster" for good mental health in the climate crisis. The concept was inspired by an experiment led by a sociologist and psychiatrist named Alexander Leighton. In the mid-twentieth century, Leighton ran a decades-long study of psychiatric problems in rural Nova Scotia, during which he followed a particularly disadvantaged community called "The Road" for more than twenty years. The 118 residents of The Road were poverty-stricken and socially isolated, didn't plan for the future, had high levels of marital strife and child neglect, lived in dilapidated houses, and experienced serious mental health problems. Other people in the county thought that residents of The Road were troublemakers who were not to be trusted.

In an attempt to help the community come together, Leighton and his colleagues prompted local government officials to offer help to the residents in the form of resources for developing the kinds of social skills that foster co-operation, alongside education and employment opportunities. Things started to gel when the residents were encouraged to agree on a goal they would like to attain as a group. They decided they wanted to raise enough money to be able to screen films in the local schoolroom, which didn't have any electricity. As they were left on their own to figure out how to do this, they learned how to lead, follow, and co-operate. Eventually the cash was raised, the schoolroom was electrified, and The Road got to enjoy their movies.

This successful pilot project was followed up with other goal-setting challenges, until the community became so skilled at co-operating and solving their own problems that the entire quality of life there had shifted. By 1963, the houses were tidy, there was less drunkenness, the community was integrated into its surrounding region, and its residents avidly took part in church, school, and other community activities. Their confidence had improved, and so too had their attitudes towards the future and the status of their mental health. Leighton concluded that co-operating to achieve shared goals was the key to making this once-impoverished community pull together and boost their own well-being.

By applying this approach to today's climate emergency, as more goals for protecting locals from its harms are communally accomplished, hope for the future will grow and emotional support will become widespread. And, as Berry says, at no point would the words "mental health" even need to be mentioned, which helps in places where the term still carries an unfortunate stigma. Then, when disaster strikes, residents will be much better equipped to co-operate and rebuild. Research has shown that this kind of community cohesiveness and social capital can even outweigh the effectiveness of economic assistance and relief from aid or government groups. Remember the shelter in Nassau after Hurricane Dorian? At first, the evacuees were 40 percent Bahamian, a proportion that quickly fell to 10 percent as Bahamians were helped out by nearby friends and family while many Haitians were left behind. The quality of our social connections has a huge impact on our well-being.

I asked Berry about Rebecca Solnit's investigation, in *A Paradise Built in Hell*, of the swelling acts of solidarity that occur amongst strangers after disaster strikes, and what that means for the "pearl in the oyster" approach. "There's lots of talk of the wonderful ways communities come together and support each other in crisis, but I think a lot of that is rhetoric more than reality," she said. "The reality is, yeah, sure, if you're in the middle of a fire or flood, anyone will give someone a hand and potentially share their box of tissues or whatever. But unless there is a lot of underlying connectedness in the community before it happens, that tends to wash away quite fast afterwards. It can be transformative in a really positive way for some communities and transformative in a negative way for others. But the evidence suggests that kind of support happens around the time of the crisis, and not long-term, unless the connections were there in the first place."

The "pearl in the oyster" approach is designed with this critical insight in mind. It's a civic-minded method of mental health innovation in the climate crisis that came up again and again in my research, even if it was called something different each time it appeared. Gary Belkin founded the Billion Minds Institute in 2019 to encourage and nourish similar mental health solutions in vulnerable communities across the United States. He believes mental health innovators and social innovators need to start seeing themselves as partners who are working on very similar things. His institute is dedicated to bolstering what he calls our "human software"—the social and emotional resilience and involvement of communities that's needed to face the climate crisis—which he calls the *social climate*. He told me, "Our human software is our biggest asset in all of this, and we don't talk about what is happening to that asset or give that asset a chance to do what it can do. Practically, I'm trying to help build it up to do what it can do."

In his role as executive deputy commissioner in the New York City Department of Health and Mental Hygiene from 2014 until 2019, Belkin became increasingly convinced that mental health providers needed to get outside clinical spaces and work with people where they are. That often meant partnering with locally trusted organizations, such as community centres and churches. He saw the need for training non-specialists in that "task-shifting" way, and considered how he might help stretch the civic muscle of front-line communities so they can better face what they're up against. As he became "personally obsessed" with the climate crisis, trying to find connections between his skills and reducing the threat, it disturbed him that nobody in policy or public discussions was talking about these easily accessible and socially embedded solutions. "What climate change will demand from us and do to us, mentally and emotionally, and what we have to do to buffer against those things, have to be these approaches," he told me.

The COVID-19 pandemic has been a scourge to people's mental health, and is representative of what we can expect to see in the climate crisis going forward, just at a much smaller scale. We now know what failing systems look like. We understand what constant urgency, a disoriented population, and immense global pressure feels like. We've seen that we are capable of massive lifestyle changes almost overnight. And in this sense, the pandemic makes it easier to understand the deep fragility of our humanity when the world tilts into some new state, no longer what it was. That's COVID-19, and it's also irreversible climate change. The pandemic widened people's appreciation of the fact that we need to learn how to cope with multiple, simultaneous global crises in practical, affordable, and community-oriented ways.

The Swiss cheese of effective leadership during the pandemic, where some nations were solid while others were gaping holes, resulted in many people dying who didn't have to. Now imagine how that will play out as climate disasters ramp up, food and water markets break down, and populations become increasingly traumatized. Many governments and institutions have proven they're not up to the task of protecting human well-being. Belkin, Berry, Patel, and a variety of others in the field of "global mental health" are demonstrating that a significant chunk of that responsibility would be more effectively managed by communities themselves, if they're given a bit of support.

The planetary health crisis is so urgent that we can't afford to go too fast, slapping familiar but deficient solutions onto the problem, instead of reimagining our approach. We must slow down enough to be thoughtful while still moving swiftly, get creative, and connect with our existential priorities as we design more capable and scalable responses. This is the only time we've got. In Belkin's words, "If the threat of extinction can't get us to put our humaneness, as a society, at the centre of what we are all about, then I guess nothing will."

The myth of the separate self has ripped many of us away from nature's intelligence and fuelled this deadly crisis. The idea that we are separate from and superior to our environment, and humanity as a whole, is a cunning and powerful lie. Deconstructing this lie, and changing the "dominant" mindset, is critical to shifting away from the path of destruction. Our personal protective equipment for the long emergency ahead depends on recognizing our interconnection. Emotional intelligence, practices to stretch the window of tolerance, the ability to sit with uncertainty, the dexterity to balance hope and fear, a commitment to not narratively foreclose the future, the belief that our actions are meaningful, an embracing of ambivalence, and a connection to a strong community that acts together to care for itself—all of these skills appear in the tool kit for surviving and thriving in the years to come. If the lion's share of mental health care moves from expensive one-on-one treatment to the "social climate," and we combine this with an ongoing faithful dedication to rapidly reduce global emissions, we still have a chance at creating a virtuous upward cycle, a fruitful tipping point—one that even fossil fuel companies and the political establishment can't thwart. The sooner we realize that no one is coming to save us, and that we must do it together ourselves, the better off we will all be.

KEY TAKEAWAYS

» Strangers often help each other out in the aftermath of disaster, but this kind of support does not necessarily stretch across social divides, nor does it stick for the long haul if strong community connections were not there in the first place.

» Focus must therefore be placed on increasing the social capital and connectedness of communities before disaster strikes, to protect mental health as more catastrophes pile up.

» Significant responsibility for mental health care can be transferred from the traditional expert-oriented, one-on-one model to a more community-based practice that puts solutions in the hands of locals (i.e., "task-shifting").

» With a bit of incentive, communities can be empowered to develop their social capital in the climate crisis, with positive effects for their mental health (i.e., "the pearl in the oyster").

» We must prioritize strengthening our "human software" and the "social climate" as we take actions to reduce global emissions and adapt—a key ingredient for protecting populations in the climate crisis, and one that hardly anyone in leadership positions is talking seriously about.

AFTERWORD

Even if I knew that tomorrow the world would go to pieces,
I would still plant my apple tree.
—MARTIN LUTHER

. . . *it is not half so important to* know *as to* feel.
—RACHEL CARSON, *The Sense of Wonder*

January 24, 2021.

Bathroom. Early morning. I say a hopeful prayer before looking at the test, despite the other minds I have. I glance at it, bring it closer to my face, and feel a tingly rush of warmth when I remark the clarity of the double blue line. I put the test in the pocket of my robe, walk into the hall, and levitate across the wooden floor until I find myself in the kitchen. I keep this secret private just long enough to make two coffees, then return to the bedroom grasping one in each hand. While my iPhone is recording on the sill above our bed, I break the news to Sebastian.

"Woah."

"Woah."

Laughter—the kind that expresses disbelief.

"I feel joy and trepidation, in that order," he says.

The word "surreal" gets mentioned several times. More laughter.

"This has been the biggest dilemma . . ."

"Of our lives. For sure. And now we've gone and done it."

It is only when Sebastian mistakenly thinks I've pressed Stop on my phone that he bursts into tears. Happy tears. Of all the moments I've captured over the years, this is my favourite recording.

Some say that those who want children will do anything to rationalize having them, even in times of societal unravelling. They may very well be right. A four-year-long internal tug-of-war didn't hold us back in the end from making this decision, despite the thousands of times I thought it would. That might be due less to the stubbornness of the human psyche when driven by deep desire and more to the specific existential meaning children have always carried. Those who yearn to be parents feel this force beating inside us, regardless of what other instincts have to say. As Kahlil Gibran writes in ***The Prophet***, published in 1923, "Your children are not your children. They are the sons and daughters of Life's longing for itself."

The act of bringing life into the world at a time when we must learn how to face ongoing catastrophic loss is indeed a dangerous dance. At the same time that Sebastian and I are learning to let go of the relative stability of the environment we've known, we must also figure out how to thrive in the face of humanity's ecological affliction. Can one thrive while living in fear? Making specific plans for the future from that mindset has only cast ever-darker shadows on my outlook, stealing all my pleasure. That's not the way I want to spend this one life I've got. Moving past the fear into some state of "acceptance" never rang as authentic either, despite how many "truth tellers" read my newsletter and email to warn me of how many years I've got left before the shit *really* hits the fan. I won't accept that there is nothing we can do to mitigate more harm and adapt compassionately along the way. I've realized that I am only as strong as my love in this world (my relationships) and for this world (my care for what happens to it), while entwined with whatever comes next.

How many mothers are migrating at this very moment with babies on their backs? How many could be alive right now but aren't because they met a violent fate? We only hear about the trauma and misery that punctuate such people's stories, and fail them further by not imagining the immense joy and meaning that also define them. A positive reflection on the emotional journey that this absurd global situation has given me is that it cracked open a far more capacious appreciation for all that shapes a life worth living. Our child's life doesn't have to look like ours for it to pass that test; plus, who knows the degree to which it still might resemble ours? Though he will never know a day without climate breakdown and ecological degradation as part of the world's baseline, we'll have to wait and see how that will inform his future dreams, abilities, and big decisions. Because of the places we were born and the families we were born into, we are part of the transition generation that has had to reckon with new expectations. The kids coming now won't have to fight their way out of society's old delusions that we had laid on us, which is radically hopeful.

Since I got pregnant, the tenor of my advocacy for climate justice has become even fiercer. The responsibility I feel for bringing another person into this situation turns out, not surprisingly, to spark a safeguarding energy that has more verve than before. To focus on our kid's well-being is a Trojan Horse for focusing even more on others' too. That is not to say it has been easy, though, since we chose a path at the fork in the road. My ambivalence, rooted in a protective mothering instinct, has crept back in from time to time. It happened most noticeably during the disaster-prone summer of 2021, while I took in headlines about surpassing tipping points, deadly heat waves, fires, and floods around the world. Wildfire smoke from northwestern Ontario and as far away as British Columbia had rolled in over Toronto, where I'd come to give birth and be around family after nearly two years of not seeing them due to the pandemic. This caused Toronto's air quality—a measure no Torontonian ever thinks of as being affected by wildfire—to plummet to some of the worst in the world, giving my hometown's sky a milky haze I'd never seen before, and making it unsafe for pregnant people to go outside for several days in a row. While firmly shut indoors, I had time to respond to journalists, who for several weeks straight had been reaching out on a nearly daily basis to ask questions about how to cope with a pervasive sense of destruction, existential angst, or it being "too late." I'd selectively do interviews while my baby played the bongos on my insides. It all felt uncanny and nightmarish.

But this shadow of mortality has consistently offered a greater realization: we'd rather have all the muck and the stress and the hardship, and all the love and the magic and the beauty—the whole human experience, which for us includes having a child—than a life where we prioritize our own preservation, operating from fear more than joy. However, I completely understand why other concerned people choose differently. I stand firmly with them and their particular forms of engagement.

There is no way to wrap this up in a bow. No clever conclusion to adequately reflect the messy complexity that comes with being awake to the world as it is—feeling the pain, anger, and heartbreak—and choosing to invest in the future anyway. I'm now in my third trimester, finalizing this book as another story begins. A child's story is always opaque to its mother at its outset. Through the blur, I can see valences of terror, but far more love, care, and connection. We know nothing yet about our son except his predicted health from a variety of tests, how much he is wanted, and his name, Atlas. Atlas—not to foist the weight of the world upon him, but with the hope that he'll have the strength of his mythological namesake to weather the challenges ahead.

ACKNOWLEDGEMENTS

Many insightful people lent me their time while I was writing this book for conversations that helped shape my thinking around the psychological underpinnings and outcomes of the planetary health crisis, as well as its various dimensions of injustice. I am very grateful to each of them, who include: Renee Lertzman, Caroline Hickman, Gary Belkin, Helen Berry, Sarah Jaquette Ray, Leslie Davenport, Panu Pihkala, Thomas Homer-Dixon, Jade Sasser, Kyle Whyte, Matthew Schneider-Mayerson, Josephine Ferorelli, Meghan Kallman, Margaret Klein Salamon, LaUra Schmidt, Aimee Lewis Reau, Paul Hoggett, Lise Van Susteren, Susan Clayton, Ashlee Cunsolo, Dennis Haseley, David Suzuki, Dan Rubin, Andrew Bryant, Robin Cooper, Rachel Ricketts, Merritt Juliano, Katie Hayes, Waubgeshig Rice, Thomas Doherty, Blythe Pepino, Emma Lim, Clover Hogan, Marcella Mulholland, Jennifer Uchendu, Cora Nally, Bethuel Isoe Nyachienga, Ceonn Edwards, Glenn Albrecht, Elizabeth Sawin, Cecilie Glerup, Travis Reider, Jennifer Atkinson, Jennifer

Mullan, Camille Parmesan, Persephone Pearl, Roy Scranton, Geoff Dembicki, Helen Caldicott, Shefali Chakrabarty, Jess Serrante, Charlie Glick, Mary Annaïse Heglar, Olúfẹ́mi Táíwò, Tamara Lindeman, Dan Suarez, Ingrid Nelson, Milanika Turner, Jaskiran Dhillon, Michael Mendez, Christie Manning, Alice Driver, Catherine Polcz, Märtha Rehnberg, Nadja Oertelt, Ashley Ahearn, Alex Trope, Tim Meijers, Jo Hamilton, Trevor Hedberg, Sami Aaron, and many more who agreed to interviews about their own personal experiences, sometimes off record.

Thank you to my brilliant colleagues in Human and Planetary Health at Stanford University and the London School of Hygiene and Tropical Medicine, as well as at Climate Cares, Imperial College London, who've supported me in researching the psychosocial impacts of the climate crisis on young people.

This book has been painful to write. The lessons in mindfulness from Plum Village, Joanna Macy's teachings, the Good Grief Network, Radical Support Collective, and climate-aware therapy have helped me maintain my sanity while thinking about the climate crisis full time for several years. Thank you also to all the readers of my newsletter *Gen Dread*, who've engaged passionately with this work. They provide proof that no one stands alone in their difficult eco-emotions, and that we can help each other live better with them.

To my book team: The always intelligent Amanda Betts from Knopf Canada has edited me with such poise and talent, I'm very grateful for her eyes on these pages. My wonderful agent Martha Webb has been a patient support and constant delight to work with. To the rest of the folks at Penguin Random House Canada who have helped polish this book—thank you. I was incredibly fortunate to be able to do research and writing residencies at TED, the Logan Nonfiction Program at the Carey Institute for Global Good, and Mesa Refuge. The space for concentration and the colleagues I gained at each residency have been huge boons to this project. My

Logan crew's feedback was especially helpful. Thank you to my friends Anders Kjemtrup and Niklas Allamand-Dib, who provided feedback on different drafts, and Farah Allamand-Dib, whose mood boards were a gift. My gratitude for my chosen family of "two countries one heart" and their common care about climate action can't be overstated.

Lastly, thank you to my family and friends who keep me grounded, and especially Sebastian, who has been a pivotal part of this book's story from the very start, and a fountain of love while I wrote it.

DISCUSSION QUESTIONS

1. Drawing from your own personal experience, have you noticed any changes in the environment around you (e.g., the neighborhood you grew up in, your favourite landscapes, vacation spots, walking trails, etc.)? What feelings come up as you explore your connection to your environment? How has your relationship with your immediate physical "non-human" environment evolved over time?

2. Explore Caroline Hickman's conceptual framework for eco-anxiety. Do you see yourself in this framework? How does your eco-anxiety manifest?

3. Discuss the similarities and differences between the environmental crisis and the COVID-19 pandemic. What does the global response to the pandemic suggest about our preparedness for the effects of climate change?

4. What are some factors that contribute to the "myth of apathy"? What are some ways we can get ourselves out of this "double bind"?

5. What is your opinion on procreation in the face of the climate crisis? Is it hopeful, brave, selfish, or uncompassionate towards children? Do you believe it is different than having kids in other eras of heightened threat? How about for communities that face life or death situations every day?

6. Describe the difference between external activism and internal activism. Why is external activism not the cure for eco-distress?

7. Do you think we're putting too much hope and pressure on younger generations to fight the fossil fuel industry and regenerate failing ecosystems? What are some meaningful ways in which we can support them for navigating their time on a hotter Earth and the dangers that come with that?

8. Describe the difference between a domination-oriented culture and one that is partnership-oriented. Explain the impact of each worldview on today's climate crisis. How do we begin to move away from domination and towards partnership? Consider Kyle Whyte's questions outlined in chapter 8: "What can I do to impact or benefit [this] story that is unfolding? What is it that my skill sets and knowledge, or things I could learn, can do to intervene in that story in a positive way?"

9. Discuss the ways in which mourning can be used to mobilize the climate movement.

10. In chapter 6, the Good Grief Network starts each meeting with the statement, "We seek to find the delicate balance between

unrealistic optimism and angry nihilism." What does unrealistic optimism look like to you? What does angry nihilism look like? What are some ways we can balance the two?

11. Following one of the activities from the Good Grief Network as outlined in chapter 6, pick up a pen and paper and write down everything you're feeling about the climate crisis. Then consider the questions: "What can I do right now in this moment? What helps me feel grounded and calm?"

12. Do you see signs of climate denial and disavowal in your life, and in the behaviour of individuals around you? Do you encounter these defences at your place of work, at school, or in your family?

13. Name some of the well-intentioned but ineffective ways of communicating about the climate crisis that impede connection and mutual understanding. What can help us engage in more dynamic and productive conversations?

14. Consider how you can expand your window of tolerance. What are some specific activities you can do?

15. To help prevent climate burnout, discuss the ways in which you personally cultivate joy in your own life. Can you activate these parts of yourself on a more regular basis, without turning away from climate reality?

16. In chapter 3, Britt Wray describes dread as "a resource floating freely in the air." In order for this generation "to capture it," we must find a "container for our overwhelming emotions." Reflect on your own "container." What does it look like and how is it intentionally created? Perhaps you have multiple.

17. What are your hopes and fears for your community, country, and the world? How do you maintain a balance between hope and fear without tipping too much to one side of the binary?

18. Explain the two parts of this sentence: "the climate crisis affects our mental health, and our mental health affects our ability to address and adapt to the climate crisis."

19. Consider the trauma inflicted by Hurricane Dorian and other climate disasters (wildfires, floods, heatwaves). Why are social capital and social connectedness vital defenses against the ways in which these extreme events negatively affect mental health?

20. After reading *Generation Dread*, what is your biggest takeaway? Have your feelings about the climate crisis changed? Have you discovered any grey areas that you'd like to explore further? Consider your experience of eco-distress and how you can harness it to create the future that you want. What are the next steps in your climate journey?

NOTES

Every quotation and reported detail that appears in the book but is not listed here was derived from an interview between the author and the subject.

INTRODUCTION

1 *It is stressful to live:* Timothy M. Lenton et al., "Climate tipping points—too risky to bet against," *Nature* 575 (November 27, 2019): 592–95, nature.com/articles/d41586-019-03595-0.

1 *It is infuriating to learn:* Naomi Oreskes and Erik M. Conway, *Merchants of Doubt* (London: Bloomsbury Press, 2010); Nathaniel Rich, *Losing Earth* (New York: Farrar, Straus and Giroux, 2019).

2 *The artists at the Bureau of Linguistical Reality:* linguistical reality (@Bureaulr), Twitter post, accessed August 22, 2021, twitter.com /Bureaulr/status/1027281631684358144.

3 *Over the last few years:* Ashlee Cunsolo et al., "Ecological grief and anxiety: The start of a healthy response to climate change?" *Lancet Planetary Health* (July 1, 2020), doi.org/10.1016/S2542-5196(20)30144-3.

3 *Tools are cropping up:* Britt Wray, "Resources for working with climate emotions," *Gen Dread* (online newsletter), accessed August 11, 2021, gendread.substack.com/p/resources-for-working-with-climate.

3 *Unfortunately, the climate crisis creates:* Brian Mastroianni, "How climate change disproportionately affects people of color," *Healthline*, April 22, 2021, healthline.com/health-news/how-climate-change-disproportionately-affects-people-of-color.

3 *Eco-anxiety researcher Panu Pihkala:* Panu Pihkala, "Climate grief: How we mourn a changing planet," *BBC Future*, April 2, 2020, bbc.com/future/article/20200402-climate-grief-mourning-loss-due-to-climate-change.

5 *Much more than just climate change:* Planetary Health Alliance, accessed June 1, 2020, planetaryhealthalliance.org/planetary-health.

5 *With each passing day:* Susan Clayton, "Environmental identity: A conceptual and an operational definition," in *Identity and the Natural Environment: The Psychological Significance of Nature*, ed. S. Clayton and S. Opotow (Cambridge, MA: MIT Press, 2003), 45–65.

5 *As we continue sucking resources:* Alfonso J. Rodriguez-Morales et al., "History is repeating itself: Probable zoonotic spillover as the cause of the 2019 novel coronavirus epidemic," *Le infezioni in medicina* 28, no. 1 (2020): 3–5. At the time of writing, I am referring to current research on the cause of the COVID-19 pandemic. However, other recent evidence points to the potential of a "lab leak," and conclusions are not yet drawn about which hypothesis is correct. See Amy Maxmen and Smriti Mallapaty, "The COVID lab-leak hypothesis: What scientists do and don't know," *Nature*, June 8, 2021, nature.com/articles/d41586-021-01529-3.

5 *Epidemiological research shows:* Raina K. Plowright et al., "Land use–induced spillover: A call to action to safeguard environmental, animal, and human health," *Lancet Planetary Health* 5, no. 1 (2021): E237–45.

8 *But then, somewhere between the UN's:* Jonathan Watts, "We have 12 years to limit climate change catastrophe, warns UN," *Guardian*, October 8, 2018, theguardian.com/environment/2018/oct/08/global-warming-must-not-exceed-15c-warns-landmark-un-report; Eric Stober, "Youth rally around the world in global climate strike," *Global News*, September 27, 2019, globalnews.ca/news/5962704/global-climate-strike-overview/; Dani Blum, "How climate anxiety is shaping family planning," *New York Times*, April 15, 2020, nytimes.com/2020/04/15/parenting/climate-change-having-kids.html.

10 *Research has shown that the kind:* Bas Verplanken et al., "On the nature of eco-anxiety: How constructive or unconstructive is habitual worry about global warming?" *Journal of Environmental Psychology* 27 (December 2020).

CHAPTER 1: THE PSYCHOTERRATIC STATE

15 *I used to think:* Gus Speth, cited from Wine and Water Watch website, accessed August 22, 2021, winewaterwatch.org/2016/05/we-scientists-dont-know-how-to-do-that-what-a-commentary.

15 *On a small farm:* Giovanni Torre, "Australia bushfires: How a 19-year-old and her father successfully defended their home from flames," *Daily Telegraph*, January 2, 2020, telegraph.co.uk/news/2020/01/02/australia-bushfires-19-year-old-father-successfully-defended/.

15 *She tried to make:* Lauren Day et al., "The stories behind the life or death moments that defined the Australian bushfire crisis," ABC News (Australia), February 3, 2020, abc.net.au/news/2020-02-03/inside-the-australian-bushfires-crisis/11890458.

16 *The MacDonells' battle:* Ibid.; Daily Telegraph, "Australian Woman Battles to Save Home from Raging Bushfires," January 3, 2020, video, 2:10, youtube.com/watch?v=Kaa9QlLCCxQ.

16 *I was on the other side:* Joel Werner and Suzannah Lyons, "The size of Australia's bushfire crisis captured in five big numbers," ABC News (Australia), March 4, 2020, abc.net.au/news/science/2020-03-05/bushfire-crisis-five-big-numbers/12007716.

19 *It is* reasonable *to get worried:* K.K. Rigaud et al., *Groundswell: Preparing for Internal Climate Migration* (Washington, DC: World Bank, 2018), openknowledge.worldbank.org/handle/10986/29461 License: CC BY 3.0 IGO; Baher Kamal, "Climate migrants might reach one billion by 2050," Inter Press Service News Agency, accessed August 20, 2021, ipsnews.net/2017/08/climate-migrants-might-reach-one-billion-by-2050.

19 *It is* appropriate *to grieve:* Jeff Tollefson, "Humans are driving one million species to extinction," *Nature* 569, no. 171 (May 6, 2019), nature.com/articles/d41586-019-01448-4.

19 *It is* logical *to be horrified:* Timothy J. Brodribb et al., "Hanging by a thread? Forests and drought," *Science* 260, no. 6488 (April 17, 2020): 261–66; Bob Berwyn, "'We need to hear these poor trees scream': Unchecked global warming means big trouble for forests," *Inside Climate News*, April 25, 2020, insideclimatenews.org/news/25042020/forest-trees-climate-change-deforestation.

19 *It is* understandable *to be scared:* Reuters, "Scientists shocked by Arctic permafrost thawing 70 years sooner than predicted," *Guardian*, June 18, 2019, theguardian.com/environment/2019/jun/18/arctic

-permafrost-canada-science-climate-crisis; Max Claypool and Brandon Miller, "Greenland's ice sheet has melted to a point of no return, according to new study," CNN.com, August 14, 2020, cnn.com/2020/08/14/weather/greenland-ice-sheet/index.html; Christopher H. Trisos, Cory Merow, and Alex L. Pigot, "The projected timing of abrupt ecological disruption from climate change," *Nature* 580 (April 8, 2020): 496–501, nature.com/articles/s41586-020-2189-9.

19 Stress *is a suitable reaction:* Chi Xu et al., "Future of the human climate niche," *Proceedings of the National Academy of Sciences* 117, no. 21 (May 2020): 11350-55, pnas.org/content/117/21/11350.

19 *It is* humane *to be gutted:* Fiona Harvey, "Polluted air killing half a million babies a year across globe," *Guardian*, October 21, 2020, theguardian.com/environment/2020/oct/21/polluted-air-killing-half-a-million-babies-a-year-across-globe.

19 *It is* fitting *to freak out:* World Meteorological Organization, *WMO Statement on the State of the Global Climate in 2019*, WMO-No 1248 (Geneva: WMO, 2020), library.wmo.int/doc_num.php?explnum_id=10211.

19 *It is* sensible *to get spooked:* Corey J.A. Bradshaw et al., "Underestimating the challenges of avoiding a ghastly future," *Frontiers in Conservation Science*, January 13, 2021, frontiersin.org/articles/10.3389/fcosc.2020.615419/full.

20 *And it is* right *to be pissed off:* Sean Illing, "It is absolutely time to panic about climate change," *Vox*, February 24, 2019, vox.com/energy-and-environment/2019/2/22/18188562/climate-change-david-wallace-wells-the-uninhabitable-earth.

20 *In* Earth Emotions*:* Glenn Albrecht, *Earth Emotions* (Ithaca, NY: Cornell University Press, 2019), 160.

20 *What Albrecht calls:* Ibid., 200.

20 *The most prevalent one:* Susan Clayton et al., *Mental Health and Our Changing Climate: Impacts, Implications, and Guidance* (Washington, DC: American Psychological Association, and ecoAmerica, 2017), 29, apa.org/news/press/releases/2017/03/mental-health-climate.pdf.

21 *For sociologist Anthony Giddens:* Anthony Giddens, *Modernity and Self-Identity* (Hoboken, NJ: John Wiley & Sons, 1991), 37.

21 Solastalgia*, another term :* Albrecht, *Earth Emotions*, 200.

21 *In light of recent:* Ibid., 71.

21 Ecological grief *is defined:* Ashlee Cunsolo and Neville R. Ellis, "Ecological grief as a mental health response to climate change–related loss," *Nature Climate Change* 8, no. 4 (2018), 275–81.

21 *Both terms relate:* Timothy Clark, "Ecological grief and anthropocene horror," *American Imago* 77, no. 1 (2020).

22 Global dread *as defined:* Albrecht, *Earth Emotions*, 199.

22 *And when I take out the trash:* Zoë Schlanger, "Race is the biggest indicator in the US of whether you live near toxic waste," *Quartz*, March 22, 2017, qz.com/939612/race-is-the-biggest-indicator-in-the-us-of-whether-you-live-near-toxic-waste/.

23 *As a result of a legacy:* Linda Villarosa, "Pollution is killing Black Americans: This community fought back," *New York Times*, July 28, 2020, nytimes.com/2020/07/28/magazine/pollution-philadelphia-black-americans.html.

23 *A 2019 survey of Americans:* Matthew Ballew et al., "Which racial/ethnic groups care most about climate change?" Yale Program on Climate Change Communication website, April 16, 2020, climatecommunication.yale.edu/publications/race-and-climate-change/.

23 *That's why, in the US:* Sally Sears, "Atlanta can claim birthplace of environmental justice movement," CBS46.com, February 25, 2021, cbs46.com/news/atlanta-can-claim-birthplace-of-environmental-justice-movement/article_4032d0c6-77c8-11eb-a545-e774faa6eb1c.html.

24 *In 2019,* Grist*:* Miyo McGinn, "2019's biggest pop-culture trend was climate anxiety," *Grist*, December 27, 2019, grist.org/politics/2019s-biggest-pop-culture-trend-was-climate-anxiety/.

24 *That same year, Oxford Languages:* Oxford Languages, "Word of the Year 2019," OUP.com, 2019, languages.oup.com/word-of-the-year/2019/.

24 *In 2020, a national poll:* Friends of the Earth, "Over two-thirds of young people experience eco-anxiety," YouGov poll, January 21, 2020, friendsoftheearth.uk/climate/over-twothirds-young-people-experience-ecoanxiety-friends-earth-launch-campaign-turn.

24 *Meanwhile, a different survey:* Marie Haaland, "Majority of young American adults say climate change influences their decision to have children," SWNS digital (website), April 20, 2020, swnsdigital.com/2020/04/majority-of-young-american-adults-say-climate-change-influences-their-decision-to-have-children/.

24 *In the same year, academics:* Susan Clayton and Bryan T. Karazsia, "Development and validation of a measure of climate change anxiety," *Journal of Environmental Psychology* 60 (June 2020).

24 *In 2020, more than half:* Jonathan Watts and Denis Campbell, "Half of child psychiatrists surveyed say patients have environmental anxiety," *Guardian*, November 20, 2020, theguardian.com/society/2020/nov/20/half-of-child-psychiatrists-surveyed-say-patients-have-environment-anxiety.

24 *The BBC surveyed:* Richard Atheron, "Climate anxiety: Survey for BBC Newsround shows children losing sleep over climate change and the environment," BBC website, March 3, 2020, bbc.co.uk/newsround/51451737.

24 *Most children responded:* Ibid.

24 *A survey conducted the same year by Våra barns klimat:* "Novusundersökning: Bara hälften av unga tror att vi löser klimatkrisen," Våra barns klimat (website), December 10, 2020, varabarnsklimat.se/novusundersokning-bara-halften-av-unga-tror-att-vi-loser-klimatkrisen.

24 *In 2021, my colleagues and I conducted:* Elizabeth Marks, Caroline Hickman, Panu Pihkala, Susan Clayton, Eric R. Lewandowski, Elouise E. Mayall, Britt Wray, Catriona Mellor, and Lise van Susteren, "Young people's voices on climate anxiety, government betrayal and moral injury: A global phenomenon," forthcoming in *Lancet Planetary Health*. Preprint available at SSRN: ssrn.com/abstract=3918955 or dx.doi.org/10.2139/ssrn.3918955.

25 *Though the stats:* "Under 40s name climate change as biggest worry," Swissinfo.ch, July 2, 2021, swissinfo.ch/eng/under-40s-name-climate-change-as-biggest-worry/46754058.

25 *As journalist Abby Higgins:* Abby Higgins, "Your climate anxiety is another person's existential crisis," *New Republic*, May 11, 2020, newrepublic.com/article/157660/climate-anxiety-another-persons-existential-crisis.

26 *Always were, for example:* Ibid.

26 *One study found that the only people:* Garret Barnwell et al., "Critical reflections from South Africa: Using the Power Threat Meaning Framework to place climate-related distress in its socio-political context," *British Psychology Society: Clinical Psychology Forum Special Issue: Psychology and the Climate and Environmental Crisis* 332 (2020).

26 *Jennifer Uchendu is:* Britt Wray, "Eco-anxiety in Nigeria," *Gen Dread* (online newsletter), April 21, 2021, gendread.substack.com/p/eco-anxiety-in-nigeria.

27 *In* Scientific American*:* Sarah Jaquette Ray, "Climate anxiety is an overwhelmingly white phenomenon," *Scientific American*, March 21, 2021, scientificamerican.com/article/the-unbearable-whiteness-of-climate-anxiety/.

27 White fragility *refers to:* Robin DiAngelo, *White Fragility: Why It's So Hard for White People to Talk About Racism* (Boston: Beacon Press, 2018).

28 *Hop Hopkins of the Sierra Club:* Hop Hopkins, "Racism is killing the planet," *Sierra*, June 8, 2020, sierraclub.org/sierra/racism-killing-planet.

28 *Ray cited the massacre:* Luke Darby, "What is eco-fascism, the ideology behind attacks in El Paso and Christchurch?" GQ.com, August 7, 2019, gq.com/story/what-is-eco-fascism.

28 *Indeed, the shooter:* Ibid.

29 *But Indigenous and anti-colonial scholars:* Heather Davis and Zoe Todd, "On the importance of a date, or, decolonizing the Anthropocene," *ACME: An International Journal for Critical Geographies* 16, no. 4 (2017): 761–80.

29 *Global warming is not only:* Erin Blakemore, "On climate change, archaeological paper digs into the effects of colonization and maltreatment," *Washington Post*, April 25, 2020, washingtonpost.com/science/on-climate-change-archaeological-paper-digs-into-the-effects-of-colonization-and-maltreatment/2020/04/24/644de4d6-84db-11ea-878a-86477a724bdb_story.html.

29 *For example, before the Industrial Revolution:* Kenneth Pomeranz, *The Great Divergence: China, Europe, and the Making of the Modern World Economy* (Princeton, NJ: Princeton University Press, 2001).

29 *European annexation of the Americas:* Ibid.

30 *This is why Kyle Whyte:* Kyle Whyte, "Is it colonial deja vu? indigenous peoples and climate injustice," *Humanities for the Environment: Integrating Knowledges, Forging New Constellations of Practice* (London: Routledge, 2016), 88–104.

30 *Many more people are now:* Jaskiran Dhillon, "Indigenous resistance, planetary dystopia, and the politics of environmental justice," *Globalizations* 18, no. 6 (2021): 898–911.

31 *Disasters that affect people directly:* Clayton et al., *Mental Health.*

31 *It can impair functioning:* Charles A. Ogunbode et al., "Negative emotions about climate change are related to insomnia symptoms

and mental health: Cross-sectional evidence from 25 countries," *Current Psychology* (2021), doi.org/10.1007/s12144-021-01385-4; Linda J. Zhang Memorial Fund for the Environment, accessed August 22, 2021, noharm.co/lindazhang/.

32 *Front-line workers who risked:* Daniel Wood, "As pandemic deaths add up, racial disparities persist—and in some cases worsen," NPR.org, September 23, 2020, npr.org/sections/health-shots/2020/09/23/914427907/as-pandemic-deaths-add-up-racial-disparities-persist-and-in-some-cases-worsen; Quentin Fottrell, "The coronavirus pandemic and Juneteenth: Black Americans are twice as likely to be hospitalized from COVID-19," *MarketWatch*, June 19, 2020, marketwatch.com/story/75-of-frontline-workers-in-new-york-the-epicenter-of-coronavirus-are-people-of-color-and-black-americans-are-twice-as-likely-to-die-from-covid-19-2020-06-01.

32 *As with climate change, young people:* Isabella Kwai and Elian Peltier, "'What's the point?' Young people's despair deepens as COVID-19 crisis drags on," *New York Times*, April 15, 2021, nytimes.com/2021/02/14/world/europe/youth-mental-health-covid.html.

32 *After four months:* Perri Klass, "Young adults' pandemic mental health risks," *New York Times*, August 24, 2020, nytimes.com/2020/08/24/well/family/young-adults-mental-health-pandemic.html.

32 *A UNICEF survey:* UNICEF Latin America and the Caribbean, "The impact of COVID-19 on the mental health of adolescents and youth," UNICEF.org, 2020, unicef.org/lac/en/impact-covid-19-mental-health-adolescents-and-youth; Marielle Wathelet et al., "Factors associated with mental health disorders among university students in France confined during the COVID-19 pandemic," *JAMA Network Open* 3, no. 10 (2020), jamanetwork.com/journals/jamanetworkopen/fullarticle/2772154.

32 *This mounting evidence:* Kwai and Peltier, "'What's the point?'"

33 *"The salvation of our world":* Martin Luther King's "Montgomery Story" address, 1955, accessed June 20, 2021, tucnak.fsv.cuni.cz/~calda/Documents/1950s/MLK_55.html.

34 *In a speech ten years later:* Karen Franklin, "Martin Luther King to psychologists: We need creative maladjustment," *Psychology Today*, January 17, 2011, psychologytoday.com/ca/blog/witness/201101/martin-luther-king-psychologists-we-need-creative-maladjustment.

34 *As a group of climate change and mental health researchers:* Cunsolo et al., "Ecological grief and anxiety."

CHAPTER 2: THE POWER OF DENIAL

36 *Humankind cannot bear:* T.S. Eliot, "Burnt Norton," *Four Quartets* (San Diego, CA: Harcourt, 1941).

36 *When we are faced by:* Paul Hoggett, "The birth of a new world (dis)order," *Climate Psychology Alliance*, January 9, 2017, climatepsychologyalliance.org/explorations/blogs/203-the-birth-of-a-new-world-dis-order.

37 *The targets are said to:* Denis Stairs, "Canada and the Cuban Missile Crisis," *The Canadian Encyclopedia* (online), February 6, 2006, thecanadianencyclopedia.ca/en/article/cuban-missile-crisis.

40 *The widespread numbing:* Robert Jay Lifton, *The Climate Swerve* (New York: New Press, 2017), 37.

41 *Timothy Morton, sometimes:* Alex Blasdel, "'A reckoning for our species': The philosopher prophet of the Anthropocene," *Guardian*, June 15, 2017, theguardian.com/world/2017/jun/15/timothy-morton-anthropocene-philosopher.

41 *It is so vast:* Timothy Morton, *Hyperobjects* (Minneapolis, MN: University of Minnesota Press, 2013).

41 *Robert Jay Lifton, with his:* Robert Jay Lifton, *Witness to an Extreme Century* (New York: Simon and Schuster, 2011).

42 *As climate-aware psychotherapist:* Paul Hoggett, introduction to *Climate Psychology: On Indifference to Disaster* (New York: Springer, 2019), 8.

42 *If we don't have a safe:* Renee Lertzman, *Environmental Melancholia* (London: Routledge, 2015).

42 *Our unprocessed anxiety:* Ibid.

43 *It comes in a few different forms:* Sally Weintrobe, *Engaging with Climate Change* (London: Routledge, 2012).

43 *Groundbreaking work by science:* Oreskes and Conway, *Merchants of Doubt*; Neela Banerjee et al., "Exxon the road not taken," *Inside Climate News*, series published 2015, accessed August 24, 2021, insideclimatenews.org/project/exxon-the-road-not-taken.

43 *Exxon, for example:* Amy Westervelt, "The Bell Labs of Energy," 2018, in *Drilled*, produced by Critical Frequency, podcast, 16:30, podcasts.apple.com/ca/podcast/the-bell-labs-of-energy/id1439735906?i=1000423823366.

43 *An internal Exxon memo:* Ibid.; Lisa Song et al., "Exxon confirmed global warming consensus in 1982 with in-house climate models," *Inside Climate News*, September 22, 2015, insideclimatenews.org

/news/22092015/exxon-confirmed-global-warming-consensus-in-1982-with-in-house-climate-models/.

43 *Science historians Naomi Oreskes:* Oreskes and Conway, *Merchants of Doubt*.

44 *In her investigative podcast:* Amy Westervelt, "Drilled, Season 3," *Critical Frequency Podcast Network*, 2020.

44 *For instance, in 2005:* Meehan Crist, "What the coronavirus means for climate change," *New York Times*, March 27, 2020, nytimes.com/2020/03/27/opinion/sunday/coronavirus-climate-change.html; Julie Doyle, "Where has all the oil gone? BP branding and the discursive elimination of climate change risk," *Culture, Environment and Eco-Politics* (Newcastle upon Tyne, UK: Cambridge Scholars Publishing, 2011).

45 *The abusive conservative backlash:* Amanda Meade, "Andrew Bolt's column mocking Greta Thunberg breached standards, press watchdog finds," *Guardian*, June 4, 2020, theguardian.com/media/2020/jun/04/andrew-bolts-column-mocking-greta-thunberg-breached-standards-press-watchdog-finds.

45 *When it comes to negating:* Joshua Ceballos, "Almost 800,000 South Florida homes could be flooded by 2100," *Miami New Times*, August 2, 2019, miaminewtimes.com/news/sea-level-rise-endangers-homes-worth-billions-of-dollars-in-south-florida-11232702.

45 *Rising sea level:* Jeff Goodell, "Miami's climate dystopia gets real," *Rolling Stone*, July 1, 2021, rollingstone.com/politics/politics-features/miami-beach-building-collapse-climate-change-1191989/.

45 *Disavowal is like:* Weintrobe, *Engaging with Climate Change*.

45 *We read that our unchecked:* Lenton et al., "Climate tipping points—too risky to bet against," *Nature* 575 (November 27, 2019): 592–95, nature.com/articles/d41586-019-03595-0.

46 *These lies then sustain:* John Steiner, *Psychic Retreats* (London: Routledge, 1993).

46 *The Oxford dictionary's: Oxford Learner's Dictionaries* (online), accessed August 24, 2021, oxfordlearnersdictionaries.com/definition/english/perverse.

46 *It's obvious that our:* Paul Hoggett, "The Psychodynamics of Climate Change Denial," UWE Bristol, September 10, 2015, video, 47:12, youtube.com/watch?v=qcn2N6kzJpo.

46 *If you're cynical:* Ibid.

47 *What psychologists call* self-efficacy*:* Mieke Garrett, *Climate Change Hopelessness* (New Plymouth, NZ: self-published, 2020).

48 *It turns out that the sour:* Sally Weintrobe, "Moral injury, the culture of uncare and the climate bubble," *Journal of Social Work Practice* 34, no. 4 (2020): 351–62.

48 *Powerful people construct bubbles:* Sally Weintrobe, *Psychological Roots of the Climate Crisis* (London: Bloomsbury Academic, 2021).

49 *"The climate bubble is far larger":* Sally Weintrobe, "Working through our feelings about the climate crisis," *The Climate Emergency: Psychoanalytic Perspectives*, The Freud Museum, May 23, 2020.

51 *Per Espen Stoknes:* Per Espen Stoknes, *What We Think About When We Try Not to Think About Global Warming* (Hartford, VT: Chelsea Green Publishing, 2015); Per Espen Stoknes, "How to transform apocalypse fatigue into action on global warming," *TEDGlobal*, September 2017, ted.com/talks/per_espen_stoknes_how_to_transform_apocalypse_fatigue_into_action_on_global_warming.

51 *We need to avoid:* Stoknes, "How to transform apocalypse fatigue."

51 *He goes so far:* Stoknes, *What We Think About*, 90.

51 *Steer clear of the apocalyptic:* Ibid., 148.

51 *As the psychologist Rick Hanson:* Rick Hanson, *Resilient* (New York: Harmony, 2018).

52 *When fear is activated:* Kasia Kozlowska et al., "Fear and the defense cascade: Clinical implications and management," *Harvard Review of Psychiatry* 23, no. 4 (2015), 263–87.

52 Executive functioning *is a kind:* Sam Goldstein and Jack A. Naglieri, *Handbook of Executive Functioning* (Berlin: Springer Science+Business Media, 2013).

CHAPTER 3: DEGREES OF ECO-DISTRESS

54 *In August of that year:* The Coquette, "On the little things," Dear Coquette (website), August 25, 2012, dearcoquette.com/on-the-little-things.

55 *When Hurricane Sandy:* Andrea Peer, "2012 Hurricane Sandy: Facts, FAQs, and how to help," WorldVision.org, September 18, 2018, worldvision.org/disaster-relief-news-stories/2012-hurricane-sandy-facts.

55 *She wrote Dear Coquette:* The Coquette, "On more fun-sized advice," Dear Coquette (website), November 1, 2018, dearcoquette.com/on-more-fun-sized-advice-56/.

56 *Consider that in 2020:* Ed Clowes, "JP Morgan warns of end to human life in climate report," *Daily Telegraph*, February 21, 2020, telegraph

.co.uk/environment/2020/02/21/jp-morgan-warns-end-human-life-leaked-climate-report/.

56 *Severe symptoms include:* Ogunbode et al., "Negative emotions about climate change"; Caroline Hickman, "We need to (find a way to) talk about . . . Eco-anxiety," *Journal of Social Work Practice* 34, no. 4 (2020): 411–24.

56 *There's also compulsive:* Panu Pihkala, "Climate anxiety," *MIELI Mental Health Finland*, 2019, https://helda.helsinki.fi/handle/10138/307626

56 *The Maldives, one of the:* Mohammed Waheed Hassan, "Climate change in the Maldives," WorldBank.org, April 6, 2010, worldbank.org/en/news/feature/2010/04/06/climate-change-in-the-maldives.

57 *Eco-anxiety: Range of feelings:* adapted from Hickman, "We need to (find a way to)," 411–24.

59 *Seminal research on the condition:* Ashlee Cunsolo, *Ecological Grief*, accessed June 4, 2021, ashleecunsolo.ca/ecological-grief.

60 *On top of that:* Ashlee Cunsolo et al., "'You can never replace the caribou': Inuit experiences of ecological grief from caribou declines," *American Imago* 77, no. 1 (2020).

60 *Research interviews with Sami:* Maria Furberg et al., "Facing the limit of resilience: Perceptions of climate change among reindeer herding Sami in Sweden," *Global Health Action* 4, no. 10 (2011).

61 *In one study of young people's:* James Diffey et al., "Young persons' psychological responses, mental health and sense of agency for the dual challenges of climate change and a global pandemic," preprint with *The Lancet,* May 26, 2021, papers.ssrn.com/sol3/papers.cfm?abstract_id=3847782.

61 *In another study, titled "My Worries":* Bas Verplanken and Deborah Roy, "My worries are rational, climate change is not," *PLoS ONE* 8, no. 9 (2013).

61 *"Instead, habitual ecological worrying"*: Ibid.

62 *It's rooted in external circumstances:* Janet Lewis et al., "Climate dialectics in psychotherapy: Holding open the space between abyss and advance," *Psychodynamic Psychiatry* 48, no. 3 (2020): 271–94.

62 *As the authors write:* Ibid.

66 *Her words punched me:* Joanna Macy, "The Hidden Promise of Our Dark Age," Bioneers, August 22, 2018, video, 31:26, youtube.com/watch?v=vzmjF1jE2K0.

69 *This gaping silence:* Kenneth J. Doka, *Disenfranchised Grief* (Hoboken, NJ: Jossey-Bass, 1989).

70 *As Panu Pihkala writes:* Panu Pihkala, "The cost of bearing witness to the environmental crisis: Vicarious traumatization and dealing with secondary traumatic stress among environmental researchers," *Social Epistemology* 34, no. 1 (2019): 86–100.

71 *"I feel hopeless, useless, futureless":* Student responses provided by Jennifer Atkinson.

72 *Fox news ran a segment:* Dan Springer, "Global warming bringing you down? Washington state college offers course on 'eco-anxiety,' 'climate grief,'" FoxNews.com, April 2, 2019, foxnews.com/us/global-warming-bringing-you-down-washington-state-offers-course-on-eco-anxiety-climate-grief.

72 *Atkinson still remembers:* Rachel Carson, *Silent Spring* (Boston: Houghton Mifflin, 1962).

73 *On the critical importance of containment:* Hoggett, introduction to *Climate Psychology*, 13.

CHAPTER 4: BABY DOOMERS

75 *The egoism of childbearing:* Sheila Heti, *Motherhood* (New York: Henry Holt, 2018).

75 *In 1988, my hometown:* Naomi Klein, *On Fire* (New York: Simon and Schuster, 2019), 177.

76 *Naomi Klein describes this:* Ibid.

76 *The UN tells us we need:* "Emissions Gap Report 2019," UN Environment Programme, November 26, 2019, unep.org/resources/emissions-gap-report-2019.

76 *But the world has already warmed:* NASA, "2020 tied for warmest year on record, NASA analysis shows," press release, January 14, 2021, nasa.gov/press-release/2020-tied-for-warmest-year-on-record-nasa-analysis-shows; Laura Millan Lombrana, "Global temperatures already 1.2°C above pre-industrial levels," Bloomberg.com, December 2, 2020, bloomberg.com/news/articles/2020-12-02/global-temperatures-already-1-2-c-above-pre-industrial-levels.

76 *As the renowned environmentalist:* Bill McKibbin, "How to combat climate depression," *New Yorker*, April 30, 2020, newyorker.com/news/annals-of-a-warming-planet/how-to-combat-climate-depression.

77 *Similarly, Caroline Hickman wrote:* Hickman, "We need to (find a way to)," 411–24.

77 *One young woman was furious:* Ibid.

80 *At that point, polls hadn't:* Marie Haaland, "Majority of young American adults say climate change influences their decision to have children," SWNS digital (website), April 20, 2020, swnsdigital.com/2020/04/majority-of-young-american-adults-say-climate-change-influences-their-decision-to-have-children/.

80 *This was before the 2019 survey:* Deborah Snow, "The big factor stopping thousands of women from starting a family," *Sydney Morning Herald*, February 11, 2019, smh.com.au/national/the-big-factor-stopping-thousands-of-women-from-starting-a-family-20190210-p50wtb.html.

80 *And it was before the poll of 1,800 American:* Claire Cain Miller, "Americans are having fewer babies. They told us why," *New York Times*, July 5, 2018, nytimes.com/2018/07/05/upshot/americans-are-having-fewer-babies-they-told-us-why.html.

81 *Matthew Schneider-Mayerson, a co-author:* Matthew Schneider-Mayerson and Leong Kit Ling, "Eco-reproductive concerns in the age of climate change," *Climatic Change* 163 (2020): 1007–23.

81 *Out of 607 respondents:* Ibid.

81 *Similarly, when I made:* Umair Ifran, "We need to talk about the ethics of having children in a warming world," *Vox*, March 11, 2019, vox.com/2019/3/11/18256166/climate-change-having-kids.

81 *Pop star Miley Cyrus:* Morgan Gstalter, "Miley Cyrus says she won't have kids because 'the earth is angry,'" *Hill*, July 12, 2019, thehill.com/blogs/in-the-know/in-the-know/452770-miley-cyrus-says-she-wont-have-kids-because-the-earth-is-angry.

81 *And Prince Harry had not yet:* Jack Guy, "Prince Harry says he is only having two children 'maximum' for the sake of the planet," CNN.com, August 2, 2019, cnn.com/2019/07/30/uk/prince-harry-babies-scli-intl/index.html.

84 *The view that human-caused:* Garrett Hardin, "Lifeboat ethics: The case against helping the poor," *Psychology Today*, September 1974, updated November 24, 2015, garretthardinsociety.org/articles/art_lifeboat_ethics_case_against_helping_poor.html.

85 *Waubgeshig Rice, an Anishinaabe:* Waubgeshig Rice, *Moon of the Crusted Snow* (Toronto: ECW Press, 2018).

86 *A lot of Black women:* American Heart Association News, "Why are black women at such high risk of dying from pregnancy complications?" American Heart Association website, February 20, 2019, heart.org/en/news/2019/02/20/why-are-black-women-at-such-high-risk-of-dying

-from-pregnancy-complications; Centers for Disease Control, "Racial and ethnic disparities continue in pregnancy-related deaths," press release, September 5, 2019, cdc.gov/media/releases/2019/p0905-racial-ethnic-disparities-pregnancy-deaths.html.

87 *Research also shows that Black and Hispanic:* Avery Ellfeldt, "Heat and racism threaten birth outcomes for women of color," *Scientific American*, June 22, 2020, scientificamerican.com/article/heat-and-racism-threaten-birth-outcomes-for-women-of-color/.

87 *In 1994, a group of:* "What is Reproductive Justice?" Sister Song Women of Color Reproductive Justice Collective, accessed August 25, 2021, sistersong.net/reproductive-justice.

87 *When she was twenty-three:* Gina Kolata, "The sad legacy of the Dalkon Shield," *New York Times*, December 6, 1987, nytimes.com/1987/12/06/magazine/the-sad-legacy-of-the-dalkon-shield.html.

87 *In 2015, climate activists Josephine Ferorelli and Meghan Kallman:* "Conceivable Future," ConceivableFuture.org, accessed August 24, 2021, conceivablefuture.org/.

88 *Kallman and Ferorelli were initially:* Center for Biological Diversity, "Men encouraged to 'get whacked for wildlife' on World Vasectomy Day," press release, November 9, 2015, biologicaldiversity.org/news/press_releases/2015/world-vasectomy-day-11-09-2015.html.

88 *American biologist Paul Ehrlich's:* Paul Ehrlich, *The Population Bomb* (New York: Ballantine, 1968).

89 *"People eating, people washing":* Ibid., 1.

89 *The next year, American activist:* Thomas Robertson, *The Malthusian Moment* (New Brunswick, NJ: Rutgers University Press, 2012), 1.

89 *In* Fatal Misconception*:* Matthew Connelly, *Fatal Misconception* (Cambridge, MA: Harvard University Press, 2008).

89 *China's one child policy:* "China to end one-child policy and allow two," BBC.com, October 29, 2015, bbc.com/news/world-asia-34665539.

89 *So did the forced sterilizations:* Marcela Valdes, "When doctors took 'family planning' into their own hands," *New York Times*, February 1, 2016, nytimes.com/2016/02/01/magazine/when-doctors-took-family-planning-into-their-own-hands.html.

89 *and several decades of surgeries:* "Forced sterilization of Indigenous women in Canada," International Justice Resource Center website, accessed August 25, 2021, ijrcenter.org/forced-sterilization-of-indigenous-women-in-canada/.

90 *Though many racist moments:* "Fertility statistics," Eurostat (website), March 2021, ec.europa.eu/eurostat/statistics-explained/index.php?title=Fertility_statistics; Sabrina Tavernise, "The U.S. birthrate has declined again. The pandemic may be accelerating the decline," *New York Times*, May 5, 2021, nytimes.com/2021/05/05/us/us-birthrate-falls-covid.html.

90 *One study showed that, on average:* Seth Wynes and Kimberly Nicholas, "The climate mitigation gap: Education and government recommendations miss the most effective individual actions," *Environmental Research Letters* 12, no. 7 (2017), iopscience.iop.org/article/10.1088/1748-9326/aa7541.

90 *Another demonstrated that a Bangladeshi:* Paul Murtaugh and Michael Schlax, "Reproduction and the carbon legacies of individuals," *Global Environmental Change* 19, no. 1 (2009): 14–20, biologicaldiversity.org/programs/population_and_sustainability/pdfs/OSUCarbonStudy.pdf.

90 *an American child, on average:* Ibid.

90 *For example, carbon emissions differ:* Rebecca Laycock Pederson and David Lam, "Second comment on 'The climate mitigation gap: Education and government recommendations miss the most effective individual actions,'" *Environmental Research Letters* 13, no. 6 (2018), iopscience.iop.org/article/10.1088/1748-9326/aac9d0/pdf.

90 *If every generation is 100 percent:* Meehan Crist, "Is it OK to have a child?" *London Review of Books* 42, no. 5 (2020).

91 *Some ethicists have used:* Travis N. Reider, *Toward a Small Family Ethic* (New York: Springer, 2016).

93 *And it did, over and over again:* Stephanie Bailey, "BirthStrike: The people refusing to have kids, because of 'the ecological crisis,'" CNN.com, June 26, 2019, cnn.com/2019/06/05/health/birthstrike-climate-change-scn-intl/index.html; Elle Hunt, "BirthStrikers: Meet the women who refuse to have children until climate change ends," *Guardian*, March 12, 2019, theguardian.com/lifeandstyle/2019/mar/12/birthstrikers-meet-the-women-who-refuse-to-have-children-until-climate-change-ends; Joseph Wulfson, "'Birth Strike' founder argues future 'too frightening' for her to have children due to climate change," FoxNews.com, March 12, 2019, foxnews.com/science/birth-strike-founder-argues-future-too-frightening-for-her-to-have-children-climate-change; Dani Blum, "How climate anxiety is shaping family planning," *New York Times*, April 15, 2020, nytimes.com/2020/04/15/parenting/climate-change-having-kids.html.

93 *They explained in a public memo:* Britt Wray, "To have a baby or not: BirthStrike changes name to evade racist affiliation," *Gen Dread* (online newsletter), September 9, 2020, gendread.substack.com/p/to-have-a-baby-or-not-birthstrike.

94 *Not long after, she was sitting:* Ontario NDP, "Horwath moves to declare a climate emergency in Ontario," press release, May 13, 2019, ontariondp.ca/news/horwath-moves-declare-climate-emergency-ontario.

94 *A few months later they launched:* Kelly McLaughlin, "Teens are pledging not to have kids until the government takes climate change seriously," *Insider*, September 18, 2019, insider.com/no-future-no-children-teens-pledge-no-kids-climate-change-2019-9.

97 *One study found that the main arguments:* Matthew Schneider-Mayerson, "The environmental politics of reproductive choices in the age of climate change," *Environmental Politics*, March 28, 2021, tandfonline.com/doi/abs/10.1080/09644016.2021.1902700

97 *The second argument:* Ibid.

97 *One child-free participant said:* Ibid.

98 *Psychiatrist Lise Van Susteren perceives:* Lise Van Susteren, "Our moral obligation," *Huffington Post*, May 25, 2011, huffpost.com/entry/our-moral-obligation_b_187751.

99 *The philosopher Anna Tsing:* Anna Tsing, *The Mushroom at the End of the World* (Princeton, NJ: Princeton University Press, 2015).

99 *"The best answer I could give":* Dougald Hine, "The sudden sense of nakedness," *Crossed Lines* (online newsletter), April 6, 2020.

99 *Feminist theorist of science and society:* Lauren O'Neill-Butler, "Interviews Donna J. Haraway," *Artforum*, September 6, 2016, artforum.com/interviews/donna-j-haraway-speaks-about-her-latest-book-63147.

99 *"Babies should be rare yet precious":* Adele Clarke and Donna Haraway, eds., *Making Kin Not Population* (Chicago: Prickly Paradigm Press, 2018), 96.

100 *Some sense a misanthropic sentiment:* Sophie Lewis, "Cthulhu plays no role for me," *Viewpoint Magazine*, May 8, 2017, viewpointmag.com/2017/05/08/cthulhu-plays-no-role-for-me/.

CHAPTER 5: STANDING IN THE SHADE OF THE CAMPHOR TREE

105 *The emotions of climate:* Britt Wray, "Betrayal, abuse, gaslighting: The story of our times," *Gen Dread* (online newsletter), May 19, 2021, gendread.substack.com/p/betrayal-abuse-gaslighting-the-story.

106 *Whereas experiencing the climate crisis:* Rodriguez-Morales et al., "History is repeating itself," 3–5. At the time of writing, I am referring to current research on the cause of the COVID-19 pandemic. However, other recent evidence points to the potential of a "lab leak" and conclusions are not yet drawn about which hypothesis is correct. See Maxmen and Mallapaty, "The COVID lab-leak hypothesis."

109 *The article was called:* Britt Wray, "Why activism isn't *really* the cure for eco-anxiety and eco-grief," *Gen Dread* (online newsletter), August 5, 2020, gendread.substack.com/p/why-activism-isnt-really-the-cure.

112 *As eco-anxiety researcher Panu Pihkala:* Pihkala, "Climate Anxiety."

113 *Research backs up what he found:* Faith Ozbay et al., "Social support and resilience to stress," *Psychiatry (Edgment)* 4, no. 5 (2007): 35–40.

117 *Psychologist Ginette Paris writes:* Ginette Paris, *Wisdom of the Psyche* (London: Routledge, 2016).

118 *It's what the environmental scholar:* Blanche Verlie, "Bearing worlds: Learning to live with climate change," *Environmental Education Research* 25, no. 5 (2019): 751–66.

119 *Rather, climate change and ecological breakdown:* Zhiwa Woodbury, "Climate trauma: Toward a new taxonomy of trauma," *Ecopsychology* 11, no. 1 (2019), liebertpub.com/doi/full/10.1089/eco.2018.0021.

119 *One useful definition:* Judith Lewis Herman, *Trauma and Recovery* (New York: Basic Books, 1992), 33.

119 *Researchers Robert Brulle and Kari Norgaard:* Robert Brulle and Kari Marie Norgaard, "Avoiding cultural trauma: Climate change and social inertia," *Environmental Politics* 28, no. 5 (2019): 886–908.

120 *As ecopsychologist Zhiwa Woodbury:* Woodbury, "Climate trauma."

120 Moral injury *was first used:* Brett Litz et al., "Moral injury and moral repair in war veterans: A preliminary model and intervention strategy," *Clinical Psychology Review* 29, no. 8 (2009): 695–706; Rita Nakashima Brock and Gabriella Lettini, *Soul Repair* (Boston: Beacon Press, 2012).

120 *Climate-aware psychoanalyst Sally Weintrobe says:* Weintrobe, "Working through our feelings."

120 *She argues that we cannot help:* Ibid.

121 *Interestingly, one study that looked at:* Samantha K. Stanley et al., "From anger to action: Differential impacts of eco-anxiety, eco-depression, and eco-anger on climate action and wellbeing," *Journal of Climate Change and Health* 1 (2021), doi.org/10.1016/j.joclim.2021.100003.

121 *Still, this study found that anger:* Ibid.

121 *Research has shown that just one hundred companies:* Carbon Disclosure Project, "New report shows just 100 companies are source of over 70% of emissions," press release, July 10, 2017, cdp.net/en/articles/media/new-report-shows-just-100-companies-are-source-of-over-70-of-emissions.

122 *Psychiatrist Robert Jay Lifton describes two kinds of guilt:* Robert Jay Lifton, *Home from the War* (London: Wildwood House, 1974).

123 *In that state, he found himself:* Daniel J. Siegel, *The Developing Mind* (New York: Guilford Press, 2012).

124 *When we only consume bad news:* Rob Hopkins, "Pre-traumatic stress disorder and the imagination with Lise van Susteren," Post Carbon Institute website, April 25, 2018, postcarbon.org/pre-traumatic-stress-disorder-and-the-imagination-with-lise-van-susteren/.

124 *"I would see a beautiful picture":* Tami Simon, "Dr. Lise Van Susteren: Emotional Inflammation: A Condition of Our Time," 2020, in *Sounds True*, podcast, 1:02:47, resources.soundstrue.com/podcast/dr-lise-van-susteren-emotional-inflammation-a-condition-of-our-time/.

125 *Pre-traumatic stress has been studied:* E. Ann Kaplan, "Is climate-related pre-traumatic stress syndrome a real condition?" *American Imago* 77, no. 1 (2020), muse.jhu.edu/article/753062.

125 *In a study that looked at Danish soldiers:* Dorthe Berntsen and David Rubin, "Pretraumatic stress reactions in soldiers deployed to Afghanistan," *Clinical Psychological Science* 3, no. 5 (2015): 663–74.

126 *American National Public Radio reported:* Nathan Rott, "Why firefighters are facing a growing mental health challenge," NPR, February 16, 2021, npr.org/2021/02/16/968252579/why-firefighters-are-facing-a-growing-mental-health-challenge.

127 *A significant body of research shows that time spent:* Qing Li, *Forest Bathing* (New York: Penguin Life, 2018).

127 *"Name it to tame it":* Dan Siegel, "Name It to Tame it," Dalai Lama Center for Peace and Education, December 8, 2014, video, 4:20, youtube.com/watch?v=ZcDLzppD4Jc.

127 *In* Emotional Resiliency*:* Leslie Davenport, *Emotional Resiliency in the Era of Climate Change* (London: Jessica Kingsley Publishers, 2017).

128 *Research shows that emotional suppression:* Nadine Andrews, "Psychosocial factors influencing the experience of sustainability professionals," *Sustainability Accounting, Management and Policy Journal* 8, no. 4 (2017), doi.org/10.1108/SAMPJ-09-2015-0080.

129 *But defending against anxiety this way:* Paul Hoggett and Rosemary Randall, "Engaging with climate change: Comparing the cultures of science and activism," in *Climate Psychology: On Indifference to Disaster* (New York: Springer, 2019), 239–61.

129 *Studies suggest that bottling up our emotions:* Yuta Katsumi and Sanda Dolcos, "Suppress to feel and remember less: Neural correlates of explicit and implicit emotional suppression on perception and memory," *Neuropsychologia* 145 (2020), 10.1016/j.neuropsycholo gia.2018.02.010; Bartosz Symonides et al., "Does the control of negative emotions influence blood pressure control and its variability?" *Blood Pressure* 23, no. 6 (2014): 323–29; Angelo Compare et al., "Emotional regulation and depression: A potential mediator between heart and mind," *Cardiovascular Psychiatry and Neurology*, 2014, doi.org/10.1155/2014/324374.

CHAPTER 6: GOOD GRIEF

132 *When it comes to thinking about my future:* Jeremie Saunders, "Embracing Your Expiry Date," TEDx Talks, December 1, 2017, video, 17:34, youtube.com/watch?v=UQniNh_6LOY.

133 *"You see, my unavoidable fate":* Ibid.

135 *Saunders says he's stopped being afraid:* Ibid.

135 *The best definition of "transformation":* "Our Strategy," Generative Somatics website, accessed August 25, 2021, generativesomatics.org /our-strategy.

136 *The psychiatrist Colin Murray Parkes:* Colin Murray Parkes, *Bereavement* (London: Routledge, 1986).

136 *The philosopher Thomas Attig:* Thomas Attig, *How We Grieve* (Oxford: Oxford University Press, 1996).

136 *As Joanna Macy writes:* Joanna Macy and Chris Johnstone, *Active Hope* (Novato, CA: New World Library, 2012), 66.

136 *The best known one comes from:* Elisabeth Kübler-Ross, *On Death and Dying* (New York: Macmillan, 1969).

137 *Climate-aware psychologist Rosemary Randall:* Rosemary Randall, "Loss and climate change: The cost of parallel narratives," *Ecopsychology* 1, no. 3 (2009): 118–29.

138 *As Randall writes:* Ibid., 119.

138 *Grief can make us come alive again:* Elizabeth Rush, "First passage," *Orion Magazine,* June 2, 2021, orionmagazine.org/article/first-passage.

142 *Becker argued in his Pulitzer Prize–winning book:* Ernest Becker, *The Denial of Death* (New York: Free Press, 1973).

145 *The first is a direct reminder of death:* Cathryn van Kessel, "Teaching the climate crisis: Existential considerations," *Journal of Curriculum Studies Research* 2, no. 1 (2020): 130.

145 *To counteract unhelpful defences:* Robert Stolorow and George Atwood, *The Power of Phenomenology* (London: Routledge, 2018), 107; Robert Stolorow, "Planet Earth: Crumbling metaphysical illusion," *Society for Humanistic Psychology* (online newsletter), July 2019, apadivisions.org/division-32/publications/newsletters/humanistic/2019/07/climate-change.

146 *In* Ecological Crisis*:* Matthew Adams, *Ecological Crisis, Sustainability and the Psychosocial Subject* (New York: Springer, 2016).

146 *A community of listeners:* Robyn Fivush, "Speaking silence: The social construction of silence in autobiographical and cultural narratives," *Memory* 18, no. 2 (2010): 88–98.

146 *Jo Hamilton completed her PhD:* Jo Hamilton, "Emotional Methodologies for Climate Change Engagement: Towards an Understanding of Emotion in Civil Society Organisation (CSO)–Public Engagements in the UK" (PhD thesis, University of Reading, UK, 2020), centaur.reading.ac.uk/95647/.

CHAPTER 7: BALANCING HOPE AND FEAR

150 *Critical thinking without hope:* Maria Popova, "Hope, Cynicism, and the Stories We Tell Ourselves," Brain Pickings (website), February 9, 2015, brainpickings.org/2015/02/09/hope-cynicism/.

150 *For several days running:* Stephanie Ip, "Number of BC households relying on air conditioners growing: BC Hydro report," *Vancouver Sun*, July 13, 2018, vancouversun.com/news/local-news/number-of-b-c-households-relying-on-air-conditioners-growing-b-c-hydro-report.

151 *It took only fifteen minutes:* "Canada Lytton: Heatwave record village overwhelmingly burned in wildfire," BBC.com, July 1, 2021, bbc.com/news/world-us-canada-57678054.

151 *More than eight hundred people died:* Andrew Weichel, "Number of deaths recorded during BC's heat wave up to 808, coroners say," CTVNews.ca, July 16, 2021, bc.ctvnews.ca/number-of-deaths-recorded-during-b-c-s-heat-wave-up-to-808-coroners-say-1.5512723.

151 *It read: "Life on Earth":* Fiona Harvey, "IPCC steps up warning on climate tipping points in leaked draft report," *Guardian*, June 23, 2021, theguardian.com/environment/2021/jun/23/climate-change-dangerous-thresholds-un-report.

151 *The climate-caused deaths:* Leyland Cecco, "'Heat dome' probably killed 1bn marine animals on Canada coast, experts say," *Guardian*, July 8, 2021, theguardian.com/environment/2021/jul/08/heat-dome-canada-pacific-northwest-animal-deaths; Paul Rogers, "How bad is this fire season in California really going to be?" *Mercury News*, July 11, 2021, mercurynews.com/2021/07/11/how-bad-is-this-fire-season-in-california-really-going-to-be/.

151 *This was quickly followed by deadly flooding:* Angela Dewan, "Germany's deadly floods were up to 9 times more likely because of climate change, study estimates," CNN.com, August 24, 2021, cnn.com/2021/08/23/europe/germany-floods-belgium-climate-change-intl/index.html; Jennifer Hassan, "Summer of floods: The climate connection between deadly downpours around the world," *Washington Post*, July 22, 2021, washingtonpost.com/world/interactive/2021/world-floods-climate/; Karan Deep Singh, "Scores die in India as monsoon rains swamp towns and send boulders tumbling," *New York Times*, July 26, 2021, nytimes.com/2021/07/26/world/asia/india-landslides-floods.html.

151 *Meanwhile, kids on TikTok:* Abbie Richards et al., "Three young climate communicators on dealing with eco-anxiety," *Euronews*, August 19, 2021, euronews.com/green/2021/08/09/exploiting-people-s-eco-anxiety-is-dangerous-it-s-up-to-climate-communicators-to-fix-thing.

152 *Then an article from the Associated Press:* Matthew Brown, "US drilling approvals increase despite Biden climate pledge," Associated Press News, July 12, 2021, apnews.com/article/joe-biden-business-science-environment-and-nature-6ac8ff49970e4b052489678b40e3ba82.

152 *As climate reporter Emily Atkin:* Emily Atkin, "'What can I do?' Anything," *Heated* (online newsletter), July 12, 2021, heated.world/p/what-can-i-do-anything.

152 *The idea of the* prospective survivor*:* Lifton, *The Climate Swerve.*

153 *In his book* The Climate Swerve*:* Ibid., 154.

153 *Hope is "such a white concept":* Nylah Burton, "People of color experience climate grief more deeply than white people," *Vice*, May 5, 2020, vice.com/en/article/v7ggqx/people-of-color-experience-climate-grief-more-deeply-than-white-people.

156 *In* Emotional Resiliency*:* Davenport, *Emotional Resiliency*, 64.

156 *Psychologist Ernst Bohlmeijer:* E.T. Bohlmeijer et al., "Narrative foreclosure in later life: Preliminary considerations for a new sensitizing concept," *Journal of Aging Studies* 25, no. 5 (2011): 364.

157 *In the words of the humanitarian:* Dougald Hine, "After we stop pretending," *Dark Mountain* 15 (August 15, 2019), dougald.nu/after-we-stop-pretending/.

158 *And though the central importance of "negative" emotions:* Nicolas M. Anspach and Gorana Draguljić, "Effective advocacy: The psychological mechanisms of environmental issue framing," *Environmental Politics* 28, no. 4 (2019), doi.org/10.1080/09644016.2019.1565468.

158 *As sociologist James Jasper writes:* James Jasper, "The emotions of protest: Affective reactive emotions in and around social movements," *Sociological Forum* 13, no. 3 (1998): 397–424.

159 *In* A Field Guide*:* Sarah Jaquette Ray, *A Field Guide to Climate Anxiety* (Berkeley, CA: University of California Press, 2020), 7.

160 *In* Man's Search for Meaning*:* Viktor Frankl, *Man's Search for Meaning* (Boston: Beacon Press, 1959).

160 *The psychoanalyst Wilfred Bion:* Wilfred Bion, *Experiences in Groups and Other Papers* (London: Routledge, 2000).

161 *"With one focus I see":* Ibid., 48.

162 *On a more political note:* Bruno Latour et al., "Anthropologists are talking—about capitalism, ecology, and apocalypse," *Journal of Anthropology* 83, no. 3 (2018): 593.

162 *The political scientist Thomas Homer-Dixon:* Thomas Homer-Dixon, *Commanding Hope* (Toronto: Knopf Canada, 2020).

163 *If we surpass a certain threshold:* Robert McSweeney, "Explainer: Nine 'tipping points' that could be triggered by climate change," *Carbon Brief*, February 10, 2020, carbonbrief.org/explainer-nine-tipping-points-that-could-be-triggered-by-climate-change.

164 *As economist Eric Beinhocker:* Eric Beinhocker, "I am a carbon abolitionist," *Democracy Journal*, June 24, 2019, democracyjournal.org/arguments/i-am-a-carbon-abolitionist/.

165 *An example of this can be framed:* John Richardson, "When the end of human civilization is your day job," *Esquire*, July 20, 2018, esquire.com/news-politics/a36228/ballad-of-the-sad-climatologists-0815/.

CHAPTER 8: THE WORLD HAS ALREADY ENDED

171 *What we pay attention to:* adrienne maree brown, "attention liberation: a commitment, a year of practice," adriennemareebrown .net, January 1, 2018, adriennemareebrown.net/2018/01/01/attention -liberation-a-commitment-a-year-of-practice/ .

172 *The gem of old growths:* Ashley Harrell, "More than six months after the fire started, Big Basin is still burning," *SFGate*, March 4, 2021, sfgate.com/california-parks/article/Big-Basin-redwoods-fire-still -burning-california-15995457.php.

172 *It was dubbed a "gigafire":* "2020 Incident Archive," CAL FIRE, accessed August 26, 2021, fire.ca.gov/incidents/2020/.

172 *For a time, San Francisco:* Alix Martichoux, "Bay Area air quality worst in the world as wildfires rage in all but one county," ABC7News .com, August 21, 2021, abc7news.com/smoke-forecast-bay-area-in-sf -today-air-pollution-wildfire/6377953/.

172 *Researchers found that exposure to:* Xiaodan Zhou et al., "Excess of COVID-19 cases and deaths due to fine particulate matter exposure during the 2020 wildfires in the United States," *Science Advances,* 7, no. 33 (2021).

173 *If the study I cited in:* Brodribb et al., "Hanging by a thread?"

174 *As Robert Jay Lifton writes:* Robert Jay Lifton, *Death in Life* (London: Weidenfeld and Nicolson, 1968), 541.

174 *Fittingly, the Greek:* Paul Corcoran, *Awaiting Apocalypse* (New York: Macmillan, 2000).

176 *But we also accidentally light:* Li Cohen, "Burst pipeline causes bubbling, steaming 'eye of fire' to emerge in the Gulf of Mexico," CBSNews.com, July 6, 2021, cbsnews.com/news/gulf-of-mexico-fire -ocean-burst-pipeline/.

176 *In* The Cybernetic Brain*:* Andrew Pickering, *The Cybernetic Brain* (Chicago: University of Chicago Press, 2010).

178 *In a* BBC *radio series:* Timothy Morton, "The End of the World Has Already Happened," BBC Radio 4, documentary, accessed August 26, 2021, bbc.co.uk/programmes/m000cl67.

178 *Cultural historian Riane Eisler:* Riane Eisler and Douglas P. Fry, *Nurturing Our Humanity* (Oxford: Oxford University Press, 2019).

178 *The Hopi, an Indigenous tribe:* Albrecht, *Earth Emotions*, 36.

179 *As the writer Charles Eisenstein:* Jem Bendell, "Deep Adaptation Q&A with Charles Eisenstein," Jem Bendell, December 14, 2019, video, 1:10:58, youtube.com/watch?v=SRYEHOStmss.

180 *We have grown to become what the mythologist:* Sharon Blackie, *The Enchanted Life* (Toronto: House of Anansi, 2018).

180 *Blackie and many others:* Plato, *Timaeus*, 30b–c, 33b.

180 *Plato's student Aristotle:* Blackie, *The Enchanted Life*.

181 *Eco-feminists have identified:* Carolyn Merchant, *The Death of Nature* (New York: Harper, 1980).

182 *She writes, "To live":* Blackie, *The Enchanted Life*, 10.

183 *The Great Unravelling:* Macy and Johnstone, *Active Hope*.

183 *The Great Turning:* Ibid.

184 *As author and activist adrienne maree brown:* brown, "attention liberation."

184 *Glenn Albrecht calls it the Symbiocene:* Albrecht, *Earth Emotions*, 103.

184 *Thomas Berry called it the Ecozoic:* Thomas Berry, "The Ecozoic Era," Schumacher Center, October 1991, centerforneweconomics.org /publications/the-ecozoic-era/.

184 *In the article "Indigenous Lessons":* Kyle Whyte et al., "Indigenous Lessons about Sustainability Are Not Just for 'All Humanity,'" in *Sustainability*, ed. Julie Sze (New York: NYU Press, 2018).

CHAPTER 9: COMMUNICATE WISELY ABOUT THE CRISIS

190 *It may sound strange:* Renee Lertzman, "How can we talk about global warming?" *Sierra*, July 19, 2017, sierraclub.org/sierra/how-can -we-talk-about-global-warming.

191 *I shared psychoanalytic research:* Weintrobe, "Moral injury," 351–62.

193 *The more emotionally neutral but all-encompassing:* Joseph Romm, "Is there a difference between global warming and climate change?" The Years Project, accessed August 26, 2021, theyearsproject.com/ask-joe /difference-global-warming-climate-change.

193 *But as economist Eric Beinhocker:* Eric Beinhocker, "I am a carbon abolitionist," *Democracy Journal*, June 24, 2019, democracyjournal.org /arguments/i-am-a-carbon-abolitionist/.

193 *Science communication researchers:* Sarah Davies and Maja Horst, *Science Communication* (New York: Springer, 2016).

193 *Another common approach:* Renee Lertzman, "Project Inside Out," ProjectInsideOut.net, accessed August 26, 2021, projectinsideout.net/.

194 *To be good partners:* Donna Haraway, *Staying with the Trouble* (Durham, NC: Duke University Press, 2016).

195 *Environmental psychologist Renee Lertzman:* Lertzman, "Project Inside Out."

196 *Some counsellors use a behaviour change technique:* Stephen Rollnick and William Richard Miller, *Motivational Interviewing* (New York: Guilford Press, 1991).

196 *As Steven Malcolm Berg-Smith:* Britt Wray, "The trick to helping people process their climate-releated dilemmas," *Gen Dread* (online newsletter), October 9, 2020, gendread.substack.com/p/the-trick-to-helping-people-process.

198 *Sixteen-year-old American climate activist:* Molly Peterson, "How to calm your climate anxiety," *New York Times*, July 23, 2021, nytimes.com/2021/07/23/well/mind/mental-health-climate-anxiety.html.

198 *As writer and activist Astra Taylor:* Astra Taylor, "Out with the old, in with the young," *New York Times*, October 18, 2019, nytimes.com/interactive/2019/10/18/opinion/old-age-president-2020.html.

199 *Global evaluations predict that kids:* Jennifer Barkin et al., "Effects of extreme weather on child mood and behavior," *Developmental Medicine and Child Neurology* 63, no. 7 (2021): 785–90.

199 *Their still-developing brains and bodies:* Ann Sanson and Susie Burke, "Climate Change and Children: An Issue of Intergenerational Justice," in *Children and Peace*, eds. Nikola Balvin and Daniel J. Christie (New York: Springer Nature, 2019).

199 *One school striker's sign put it this way:* Jason Plautz, "The environmental burden of Generation Z," *Washington Post*, February 3, 2020, washingtonpost.com/magazine/2020/02/03/eco-anxiety-is-overwhelming-kids-wheres-line-between-education-alarmism/.

199 *How could they not be expressing:* Helen Clark et al., "A future for the world's children? A WHO–UNICEF–*Lancet* Commission," *Lancet* 396, no. 10223 (2020): 605–58; World Health Organization, "World failing to provide children with a healthy life and a climate fit for their future: WHO–UNICEF–*Lancet*," press release, February 19, 2020, who.int/news/item/19-02-2020-world-failing-to-provide-children-with-a-healthy-life-and-a-climate-fit-for-their-future-who-unicef-lancet.

199 *One study found that between 2009 and 2017:* Jean Twenge et al., "Age, period, and cohort trends in mood disorder indicators and suicide related outcomes in a nationally representative dataset, 2005–2017," *Journal of Abnormal Psychology* 128, no. 3 (2019): 185–99.

200 *"We need to create more spaces":* Maria Ojala, "Eco-anxiety," *RSA Journal* 164, no. 4 (2018): 15.

200 *As World War II veteran:* Howard Kassinove and Raymond Tafrate, *Anger Management* (Oakland, CA: Impact Publishers, 2002), 244.

200 *Child psychologist Jo McAndrews*: Jo McAndrews, "Supporting Children in the Face of Climate Change," Jo McAndrews, November 29, 2018, video, 1:29:49, youtube.com/watch?v=2bm18_G4n2Y.

CHAPTER 10: THE POTENCY OF PUBLIC MOURNING

203 *The primary emotional work we need to do:* Shierry Weber Nicholsen, *The Love of Nature and the End of the World* (Cambridge, MA: MIT Press, 2003), 140.

204 *Greta Thunberg and German climate activist:* Reuters, "QUOTEBOX—'Neoliberalism is a death cult': Quotes from climate activists in Madrid," Reuters.com, December 2, 2019, reuters.com/article/climate-change-accord-greta/quotebox-neoliberalism-is-a-death-cult-quotes-from-climate-activists-in-madrid-idUSL8N28J2NX; COP25, "Press Conference at COP25 by Greta Thunberg, Luisa Neubauer and Fridays for Future," United Nations Framework Convention on Climate Change, September 12, 2019, video, 38:20, unfccc-sb50.streamworld.de/videos/press-conference-greta-thunberg-luisa-neubauer-and-fridays-future.

204 *A young man, Carlon Zackhras:* Ibid.

206 *In the early 1980s, a menacing virus*: Kaiser Family Foundation, "The HIV/AIDS Epidemic in the United States: The Basics," fact sheet, June 7, 2021, kff.org/hivaids/fact-sheet/the-hivaids-epidemic-in-the-united-states-the-basics/.

206 *Despite the coincidence:* Maria La Ganga, "The first lady who looked away: Nancy and the Reagans' troubling Aids legacy," *Guardian*, March 11, 2016, theguardian.com/us-news/2016/mar/11/nancy-ronald-reagan-aids-crisis-first-lady-legacy.

206 *At that point, more than five thousand:* Walt Odets, "Ronald Reagan presided over 89,343 deaths to AIDS and did nothing," *Literary Hub*, July 22, 2019, lithub.com/ronald-reagan-presided-over-89343-deaths-to-aids-and-did-nothing/.

207 *As the AIDS memorial website reads:* "The AIDS quilt, learn more," National AIDS Memorial, accessed March 1, 2020, aidsmemorial.org/theaidsquilt-learnmore.

207 *This process would not have succeeded:* "Climate Change as the Work of Mourning," *Mourning Nature*, eds. Ashlee Cunsolo and Karen E. Landman (Montreal: McGill-Queen's University Press, 2017), 176.

210 *Besides the great auk:* Jeremy Hance, "Why don't we grieve for extinct species?" *Guardian*, November 19, 2016, theguardian.com

/environment/radical-conservation/2016/nov/19/extinction-remembrance-day-theatre-ritual-thylacine-grief.

211 *In a blog post titled:* Persephone Pearl, Rachel Porter, and Emily Laurens, "On racism and environmentalist practice—reflections on a journey," *Remembrance Day for Lost Species* (blog), October 22, 2019, lostspeciesday.org/?p=1458.

212 *Thomas Van Dooren, a philosopher of extinction:* Thomas Van Dooren, "Pain of extinction: The death of a vulture," *Cultural Studies Review* 16, no. 2 (2010): 273.

212 *The process is accelerated by:* Bachi Karkaria, "Death in the city: How a lack of vultures threatens Mumbai's 'Towers of Silence,'" *Guardian*, January 26, 2015, theguardian.com/cities/2015/jan/26/death-city-lack-vultures-threatens-mumbai-towers-of-silence.

CHAPTER 11: STRONGER COMMUNITIES FOR A BETTER FUTURE

216 *We've got to be as clear-headed:* James Baldwin and Margaret Mead, *A Rap on Race* (London: Michael Joseph, 1971).

216 *On September 1, 2019:* "Hurricane Dorian six-month report," Americares, March 2020, americares.org/wp-content/uploads/Americares-Hurricane-Dorian-Six-month-Report.pdf.

216 *In the wee hours of the next morning:* Associated Press, "Nearly 2 weeks after Hurricane Dorian, 1,300 people still missing in the Bahamas," ABCActionNews.com, September 13, 2019, abcactionnews.com/weather/hurricane/nearly-2-weeks-after-hurricane-dorian-1-300-people-still-missing-in-the-bahamas.

216 *Seventy thousand people were displaced:* "The Dorian Disaster," Americares, August 28, 2019, americares.org/emergency-program/hurricane-dorian/.

216 *Airports were completely submerged:* Alfred Sears, "Hurricane Dorian: A reset, pt. 1," *Nassau Guardian*, December 4, 2019, thenassauguardian.com/hurricane-dorian-a-reset-pt-1; "Bahamas—South Riding Point," Equinor, accessed August 26, 2021, equinor.com/en/what-we-do/terminals-and-refineries/bahamas.html.

217 *The economic losses were estimated:* Inter-American Development Bank, "Damages and other impacts on Bahamas by Hurricane Dorian estimated at $3.4 billion: report," press release, November 15, 2019, iadb.org/en/news/damages-and-other-impacts-bahamas-hurricane-dorian-estimated-34-billion-report.

217 *The Inter-American Development Bank:* Ibid.

217 *American officials described:* Stuart Ramsay, "Hurricane Dorian: Nobody knows how many lived—and died—at Ground Zero," *Sky News*, September 9, 2019, news.sky.com/story/hurricane-dorian-nobody-knows-how-many-lived-and-died-at-ground-zero-11805362.

218 *Disaster doesn't sort us out:* Rebecca Solnit, *A Paradise Built in Hell* (New York: Penguin Books, 2010), 7.

218 *We learn the same lesson:* Klein, *On Fire*, 215.

219 *The initial group of 1,200 evacuees:* "The Dorian Disaster," Americares.

220 *The status of the residents was mixed:* Alfred Sears, "Hurricane Dorian: A reset, part 2," *Nassau Guardian*, December 10, 2019, thenassauguardian.com/hurricane-dorian-a-reset-part-2/.

220 *It was strikingly clear:* Ramsay, "Hurricane Dorian: Nobody knows."

220 *While the official loss of life:* Sears, "Hurricane Dorian: A reset, pt. 1."; Brianna Sacks, "The grim, exhausting task of finding those killed by Hurricane Dorian takes its toll," *BuzzFeed News*, September 15, 2019, buzzfeednews.com/article/briannasacks/cadaver-dogs-hurricane-dorian-victims-search.

220 *Rescue dogs:* Ibid.

221 *The International Organization for Migration was tracking:* IOM UN Migration, "IOM tracks repatriations of Haitian migrants from the Bahamas," press release, November 15, 2019, iom.int/news/iom-tracks-repatriations-haitian-migrants-bahamas.

221 *He worked in Liberia:* "Chain-Free Initiative," Mental Health Innovation Network, mhinnovation.net/innovations/chain-free-initiative.

222 *When Dorian struck, it was the third:* Associated Press, "A look at previous hurricanes that have affected the Bahamas," APNews.com, August 31, 2019, apnews.com/article/bahamas-caribbean-latin-america-hurricane-dorian-hurricanes-0e4771697208445db7a05ebfeda1b1b5.

222 *As global temperatures rise, more heat energy:* "What's the difference between hurricanes, cyclones and typhoons?" BBC website, August 10, 2019, bbc.co.uk/newsround/24879162.

223 *So as the world gets hotter:* Jeff Berardelli, "How climate change is making hurricanes more dangerous," *Yale Climate Connections*, July 8, 2019, yaleclimateconnections.org/2019/07/how-climate-change-is-making-hurricanes-more-dangerous/.

223 *Women migrating under the threat:* Vikram Patel et al., "The *Lancet* Commission on global mental health and sustainable development," *Lancet* 392, no. 10157 (2018): 16.

223 *Farmers who remain and try:* Tamma A. Carleton, "Crop-damaging temperatures increase suicide rates in India," *Proceedings of the National Academy of Sciences of the United States of America* 114, no. 33 (2017): 8746–51.

223 *When the next disaster violently upends:* R.C. Kessler et al., "Trends in mental illness and suicidality after Hurricane Katrina," *Molecular Psychiatry* 13, no. 4 (2008): 378–84; Heather Keenan et al., "Increased incidence of inflicted traumatic brain injury in children after a natural disaster," *American Journal of Preventative Medicine* 23, no. 6 (2004): 189–93; Jessica Fritze et al., "Hope, despair and transformation: Climate change and the promotion of mental health and wellbeing," *International Journal of Mental Health Systems* 2, no. 13 (2008).

223 *Poor people's property:* Susan Clemens et al., "Summer of sorrow: Measuring exposure to and impacts of trauma after Queensland's natural disasters of 2010–2011," *Medical Journal of Australia* 199, no. 8 (2013): 552–55.

223 *Anyone who didn't get adequately warned:* Laura Anderko et al., "Climate changes reproductive and children's health: A review of risks, exposures, and impacts," *Pediatric Research* 87 (2020): 414–19.

224 *People who need life-saving medications:* Sae Ochi et al., "Disaster-driven evacuation and medication loss: A systematic literature review," *PLOS Currents* 18, no. 6 (2014).

224 *In heat waves, people who live alone:* Eric Kleinenberg, *Heat Wave* (Chicago: University of Chicago Press, 2003).

224 *Some antipsychotic drugs:* Lynette Cusack, "Heatwaves and their impact on people with alcohol, drug and mental health conditions: A discussion paper on clinical practice considerations," *Journal of Advanced Nursing* 67, no. 4 (2011): 915–22.

224 *Murders, assaults, suicides, and hospitalizations for self-harm:* C.A. Anderson, "Heat and violence," *Current Directions in Psychological Science* 10, no. 1 (2001): 33–38; C.A. Anderson and M. DeLisi, "Implications of Global Climate Change for Violence in Developed and Developing Countries," in *Psychology of Social Conflict and Aggression*, eds. J. Forgas et al. (Hove, UK: Psychology Press, 2011), 249–65; C.A. Anderson et al., "Temperature and aggression," *Advances in Experimental Social Psychology* 32 (2000): 63–133; Marshall Burke et al., "Higher temperatures increase suicide rates in the United States and Mexico," *Nature Climate Change* 8 (2018): 723–29.

224 *In rich regions of the world:* Helen Berry et al., "The case for systems

thinking about climate change and mental health," *Nature Climate Change* 8 (2018): 282–90.

225 *Countries with right-leaning governments:* T. Blakely and S. Collings, "Is there a causal association between suicide rates and the political leanings of government?" *Journal of Epidemiology and Community Health* 56 (2002): 722; A. Page et al., "Suicide and political regime in New South Wales and Australia during the 20th century," *Journal of Epidemiology and Community Health* 56, no. 10 (2002): 766–72; M. Shaw et al., "Mortality and political climate: How suicide rates have risen during periods of Conservative government, 1901–2000," *Journal of Epidemiology and Community Health* 56 (2002): 723–25; Sarah Boseley, "Suicide rates tend to rise under Tory rule," *Guardian*, September 19, 2002, theguardian.com/politics/2002/sep/19/uk.conservatives.

226 *Right now, in regions of the world:* Theresa Hoeft et al., "Task-sharing approaches to improve mental health care in rural and other low-resource settings: A systematic review," *Journal of Rural Health* 34, no. 1 (2018): 48–62; D. Javadi et al., "Applying systems thinking to task shifting for mental health using lay providers: A review of the evidence," *Global Mental Health* 4, no. E14 (2017).

226 *Vikram Patel, a psychiatrist:* Vikram Patel, "Mental Health for All by Involving All," TEDGlobal2012, June 2012, video, 12:06, ted.com/talks/vikram_patel_mental_health_for_all_by_involving_all?language=en.

226 *In one clinical trial that Patel led*: Ibid.

226 *He says that psychosocial support skills:* Ibid.

226 *At this moment when more people live alone:* Esteban Ortiz-Ospina, "The rise of living alone: How one person households are becoming increasingly common around the world, " *Our World in Data*, December 10, 2019, ourworldindata.org/living-alone; John Cacioppo and Stephanie Cacioppo, "The growing problem of loneliness," *Lancet* 391, no. 10119 (2018): 426.

227 *The potential of social relationships:* Katie Hayes et al., "Factors influencing the mental health consequences of climate change in Canada," *International Journal of Environmental Research and Public Health* 16, no. 9 (2019): 1583.

227 *One of Helen Berry's studies:* Brett McDermott et al., "Vulnerability factors for disaster-induced child post-traumatic stress disorder: The case for low family resilience and previous mental illness,"

Australian and New Zealand Journal of Psychiatry 4, no. 4 (2010): 384–89.

227 *This could be because social connectedness:* Hayes et al., "Factors influencing the mental health consequences," 1583; Daniel P. Aldrich, *Building Resilience* (Chicago: University of Chicago Press, 2012).

228 *Berry calls this approach:* Helen Berry, "Pearl in the oyster: Climate change as a mental health opportunity," *Australasian Psychiatry* 17, no. 6 (2009): 453–56.

228 *In the mid-twentieth century:* John Schwab, *Sociocultural Roots of Mental Illness* (New York: Springer, 1978).

228 *The 118 residents of The Road:* D.C. Leighton et al., "Psychiatric findings of the Stirling County Study," *American Journal of Psychiatry* 119, no. 11 (1963): 1021–26.

228 *Things started to gel:* A.H. Leighton, "Poverty and social change," *Scientific American* 212, no. 5 (1965): 21–27.

229 *By 1963, the houses were tidy:* Ibid.; A.H. Leighton, "Some notes on preventative psychiatry," *Canadian Journal of Psychiatry* 12 (1967): 43–50.

229 *Leighton concluded that co-operating:* A.H. Leighton, "Social science and psychiatric epidemiology: A difficult relationship," *Acta Psychiatrica Scandinavica Supplementum* 385 (1994): 7–12.

229 *Research has shown that this kind of community cohesiveness:* Mary LaLone, "Neighbors helping neighbors: An examination of the social capital mobilization process for community resilience to environmental disasters," *Journal of Applied Social Science* 6, no. 2 (2012): 209–37.

AFTERWORD

234 *Even if I knew that tomorrow:* Martin Luther. There is an inconclusive debate around the origins of this quote, with many references attributing it to the German theologian Martin Luther despite the lack of evidence from his writings. See Scott Hendrix, *Martin Luther* (Oxford: Oxford University Press, 2010).

234 *... it is not half so important to* know *as to* feel*:* Rachel Carson, *The Sense of Wonder* (New York: HarperCollins, 2017), 21.

237 *This caused Toronto's air quality:* Anya Zoledziowski, "Inspiring! Canada's air quality is some of the worst in the world," *Vice*, July 20, 2021, vice.com/en/article/wx57qx/inspiring-canadas-air-quality-is-some-of-the-worst-in-the-world.

INDEX

BRITT WRAY is a writer and broadcaster researching the emotional and psychological impacts of the climate crisis. Born and raised in Toronto, Canada, she is a postdoctoral fellow at Stanford University and the London School of Hygiene & Tropical Medicine, where she investigates the mental health consequences of ecological disruption. She holds a PhD in science communication from the University of Copenhagen. She is the author of *Rise of the Necrofauna: The Science, Ethics and Risks of De-Extinction*. Her work has been featured in the *New York Times*, *Washington Post*, *Guardian*, and *Globe and Mail*, among other publications. She has hosted several podcasts, radio and TV programs with the BBC and CBC, is a TED Resident, and writes *Gen Dread*, a newsletter about staying sane in the climate crisis: gendread.substack.com.